'GrowerTalks'®

ON PLUGS

Geo. J. Ball
PUBLISHING

GrowerTalks® on Plugs

Library of Congress Cataloging in Publication Data

GrowerTalks® on Plugs.

90-082395

Includes index.

ISBN 0-9626796-0-7

Printed in the United States of America.

Editor: Debbie Hamrick
Copy editor: Marie Hope
Production manager: Janet E. Sandburg
Cover design and illustration: John N. Landahl
Printed by Palmer Publications.

CONTENTS

Introduction

Plugs, they have and are continuing to change the face of North American floriculture and they're sweeping into Europe and Asia too. Plugs are efficient, they facilitate mechanization, they provide a uniform start to any seed grown crop and they increase the maximum number of turns you can obtain from valuable growing space. Over the brief course of the rise of plug production since Kube-Pak in New Jersey pioneered seedlings in media, U.S. and Canadian growers and researchers have learned much about how to and how not to grow plugs. The development of plug technology has been reported in *GrowerTalks* over the years.

GrowerTalks on Plugs is a collection of the best technical articles on how to grow plugs that have appeared in the past 60 issues of *GrowerTalks* magazine. Within the following pages, you'll find specific information on seed germination, in-depth profiles of some of North America's best plug growers and details on how to recognize and solve production problems.

If you are a grower or user of plugs, you are a part of one of the most exciting developments in floriculture. Congratulations, and may your future be successful and prosperous.

The Editorial Staff of *GrowerTalks* magazine

General Information

Industry Talks

December 1989

based on an interview by Julie A. Martens

"In time, it won't make any more sense for finishers to produce their own plugs than it would to produce their own seed or flats." Gene Greiling

Name: *Gene Greiling*
Occupation: *President, Greiling Farms, Inc.*
(doing business as Natural Beauty of Florida)
    Background: *Gene was born into the family produce business in Wisconsin. Gene and his wife Arlis, started a greenhouse range in 1971 at Denmark, Wisconsin, which has expanded to 14 acres. In 1979 they moved to Florida where they now live. The Florida greenhouse range has grown to 27 acres and is known as Natural Beauty of Florida. Their latest range is in the mountains of North Georgia and is approximately 8 acres.*
    All three locations produce plugs, foliage and poinsettias and finish bedding plants. Spark Plugs and Expeditors are marketed through the Ball Seed Co. and finished products are marketed direct by Greiling Farms Inc. to chain stores, garden centers and landscapers. Gene and his team of managers make good use of the company airplane in moving between the three locations and their customer base, which is nationwide.

"The impact of plugs on the bedding plant industry has created far reaching changes in several areas. Where before there was the seed company and the traditional bedding plant grower, there is now evolving the plug specialist and the bedding plant finisher. These two newcomers to the industry are causing the seed companies and the traditional bedding plant growers to rethink how their products are produced and marketed.

"The large volume plug specialist is a new type of customer that the seed companies have to adjust their marketing efforts to accommodate. Plug production requires that seed quality and vitality standards be rewritten by the seed producers. Seed companies that fail to respond to the needs of the plug specialist will find themselves unable to sell to a larger and larger percent of the market."

Buying plugs brings better profits

"The traditional bedding plant producer now has a choice of seed or plug—and increasingly after doing his cost accounting is finding that buying plugs and not seeds is the way to better profits. It will also be proven that the smallest and cheapest plugs are not the way to produce the best and most profitable finished bedding plant crops.

"Larger plugs finish more uniformly in shorter grow times with less talent required and in less expensive structures. Larger plugs also allow for more crop turns per season. The smallest plug is not the way to the cheapest finished flat. The bedding plant producer should weigh the **value** of the plug, not just the cost, in deciding what size plug is best for him.

"The bedding plant finisher who starts with a well grown plug does not need many of the technical skills or expensive facilities to produce a good crop of annuals. This creates the opportunity to produce and market a crop of seasonal bedding plants without expensive facilities and technical experience.

"In short, the plug has made it easy and inexpensive to get into the bedding plant business. Bedding plant finishers can increase their production capacity by double cropping their existing greenhouse space or by expanding with very inexpensive additional space and adding unskilled labor. The plug makes all of this possible."

Mechanized transplanting and streamlining transportation

"The biggest challenge in plug production is in dealing with the peak demand and the seasonality of the business. The most important developments in plug production technology are in the area of seed priming and pre-germing seed to allow for 100 percent stands of uniform plugs. This technology allows both lower production and shipping costs and facilitates mechanical transplanting.

"The large volume finisher will demand plugs produced in a tray or system that will be adaptable to mechanical transplanting. This will require 100 percent stands of very uniform plugs. The delivery of these plugs from producer to finisher must be done in an efficient manner without compromising the quality of the plugs.

"Transporting plugs from the plug specialists to the bedding plant finisher is still an area that needs improvement. The successful plug producer will be the one that accepts the responsibility of getting his product to his customers quickly and in an inexpensive manner.

"Expensive packaging and delivery methods must and will be replaced by better and more efficient means. This will lead to higher quality and lower priced plugs delivered to the finisher. The plant industry in this country is in serious need of a temperature controlled, cart oriented trucking service, much like the one used in Europe.

"When this is accomplished, it will make very little economic sense for a bedding plant finisher to do his own plugs. In time, it will make no more sense for finishers to produce their own plugs than it would be to produce their own seed or plastic flats."

"

Julie A. Martens is managing editor of GrowerTalks *magazine.*

Research Update

December 1989

by David R. Dreesen and Robert W. Langhans
Cornell University, Ithaca, New York

"We can predict to the day when a certain size plug can be obtained."

Controlled environment growth rooms haven't been widely used for flower production. To be profitable, a crop must meet several criteria: high crop value per unit of production area, short crop time in the room, and large crop response to the environmental conditions in the growth room.

Bedding plant plugs have high crop values ($10 to $12 per square foot) and short crop times (two to eight weeks in the greenhouse, depending on species). Looking at temperature, light and CO_2 concentration effects on plug growth, we learned that, if all conditions in the growth rooms are uniform crop to crop—including nutrition—we can predict to the day when a certain size plug can be obtained.

Growth rooms used in the research are 8 feet wide by 11 feet long and 7 feet high. Ebb and flow benches allow for subirrigation and water recycling, and the tray support system air prunes roots. Cooling is via conductive heat loss to the greenhouse and exhaust fans (when necessary).

Choosing the optimum growth environment depends on what you're putting in

Growing plugs in a controlled environment makes it possible to have uniform and predictable growth responses. Based on growing conditions we used in our research with Accent Red impatiens, we're able to make some recommendations about growing plugs in growth chambers. Here are some tips that you can put to work in your production programs.

• **Temperature.** Increasing air temperatures increases leaf area, which almost doubles as 24-hour average daily temperatures go from 64 to 74 degrees F. Seedling height at Day 10 (cotyledonary stage) is 30 percent greater at 84 versus 64 degrees F; at Day 19 (marketable stage), high-temperature seedlings are up to 50 percent taller. With 24-hour lighting at 1,700 to 2,600 footcandles, optimum plug media temperature is 75 to 80 degrees F.

• **Light.** By Day 19, seedlings receiving 3,500 fc of light had three times as much dry weight as low light seedlings (700 fc). High light causes much faster growth, resulting in tall but stocky seedlings; low light causes stretching. To get stocky, high quality plugs requires at least 1,500 fc of light.

• **CO_2 levels.** A 46 percent growth increase occurs at 1,100 parts per million CO_2 with 2,300 fc of light. With both higher and lower CO_2 levels, growth is less. At low light levels (1,500 fc), adding CO_2 doesn't increase growth—high CO_2 concentrations can even depress plug growth. Some leaf curling, chlorosis and other abnormalities occur with high CO_2, especially as plugs approach marketable stage. If you plan to use CO_2 enrichment with your plugs, at least 2,000 fc of light are necessary to see a boost in seedling growth.

• **Putting the factors together.** As we looked at each environmental factor individually, we came up with optimum growing conditions: 74 degrees F, 1,000 ppm CO_2 and 4,500 fc. In this environment we get a seedling that's two (by Day 10) to three (by Day 19) times larger than plugs in a "baseline" or typical growth environment (64/73 degrees F day/night, ambient CO_2 and 2,000 fc). Decreasing light levels to 2,800 fc doubles seedling growth by Day 19 when compared to plugs under baseline conditions.

Under our optimum conditions, we can produce a marketable seedling in 15 to 16 days. Within this 16-day growth period, the first four days are the germination period—much lower light and ambient CO_2. This means that the costly high light environment is needed for only 12 days to produce a high quality, stocky seedling. Our near optimum environment—74 degrees F, 1,000 ppm CO_2 and 2,800 fc—yields similar growth increases for Super Elfin Blush impatiens, Ringo Scarlet geraniums and Prelude Rose fibrous begonia.

What's the bottom line on a growth chamber with the "optimum" environment?

We've developed a conceptual design of a controlled environment plug facility with 11,000 square feet of production area. The lighting system is designed to have an initial light intensity of 3,000 fc, which will drop to 2,200 fc for year-old lamps (sufficient light to continue to get a large growth increase from CO_2 enrichment at 1,000 ppm).

An optimistic scenario assumes three-week crop time in the optimum high light environment with an electricity cost of 5 cents per kilowatt-hour; a pessimistic scenario, a 3¹/₂-week crop time and 10-cent-per-kilowatt-hour electricity. Capital costs of the facility are estimated at $800,000; operating costs range from $1.4 million to $1.8 million, assuming 5 cents per plug and 75 to 80 percent seedling stands.

With cheap electricity, a potential profit of $400,000 occurs under the best scenario; a large loss ($300,000) under the worst scenario. In addition to cheap electricity, a cool climate is a site requirement to reduce ventilation needs and maximize CO_2 enrichment periods. Using Ithaca, New York, weather, we estimate that CO_2 enrichment would be feasible for 80 percent of the hours in a typical year. The necessity for cheap electricity points to locating near cheap hydroelectric power, using off-peak rates to the maximum possible, or generating electric power on site using cogeneration.

This economic analysis assumes that year-round production is feasible. Bedding plant plugs could be produced in the late winter and spring, perennial plugs in summer, and cut flower, winter pot crop and shrub or tree plugs in fall and early winter. We grew perennial ground covers in summer using the growth rooms. These plugs could be sold directly to landscapers for fall planting or to growers for overwintering. We found 1-inch square plugs (128 tray) of these groundcovers could be transplanted to the landscape in late summer and become well established by fall even during low rainfall periods.

„

David Dreesen is a postdoctoral associate and Bob Langhans is professor of floriculture, Cornell University, Ithaca, New York. Equipment and supplies donated by Harris Moran Seed Co., Ivy Acres Inc., Ball Seed Co., BFG Supply Co., Vaughan's Seed Co., H.G. German Seed Co. and W.R. Grace Co.

The 3¢ bedding plant plug

December 1989

by Debbie Hamrick

800 plugs. They're at last an economic means to put plugs in flats, yielding the highest density so far available on the plug bench at 300 per square foot. There's a major warning, though: When 800 plugs are ready to transplant, they cannot be held. Will 800s make major inroads into the industry?

Controversial could be one word to describe 800-cell plugs. Ask a half-dozen people and you are likely to hear comments ranging from "everyone used to think 600s were too small . . . in an 800 you're planting basically a germinated seed," and "the margin of error is about 2.7 times worse than 512s" to "with 800s growers can now afford to put a plug in the flat."

The bottom line on the controversy: 800 plugs are small, and they cannot be held. They cannot be held for a few days and certainly not for a week, whether it's on the plug bench waiting for free finishing space or beside the transplanting line after arrival.

The reason: 800 cells contain 1.5 cubic centimeters of media in a ⁵/₈-inch deep column, slightly more than 648s. The goal with plugs is to maintain an actively growing root system. This is a difficult job with such a small area in which to work.

"In an 800, the root ball forms quicker, filling the cell quickly. As you get faster rootbounding, you get crowding and root hair inactivity. Nutrients aren't taken up as quickly and plants end up smaller," says Dave Koranski, Iowa State University, Ames. "With more media volume such as a 400, there's a more active root system with slower rootbounding. Finished plants produce more axillary shoots and flowers. If you transplant 800s when they are ready, it's O.K. If you wait, you're in trouble."

There's no margin for error

One thing everyone agrees on is that 800 plug trays are not for beginning plug growers. Their small size allows no flexibility and no buffer for error. Specialist plug growers and larger bedding plant growers who have made significant investments in the environment in their plug ranges are more likely to have success. The reason growing in 800 cells is so tricky, Dave says, is because of the cell's shallow column. Because the column is so shallow, 5/8-inch, water management is difficult.

If 800s are so difficult to grow, then what's all the hoopla about? Simply because when an 800 cell is grown well, you can't tell the difference in 800 or 400 plants above the media line. Finishing times in many cases are later. Some varieties finish the same, others finish one to one and a half weeks later.

How much media volume?*	
Cells/plug tray	Volume (cubic centimeters)
800 square	1.5
648 square	1.4
406 round	3.5
406 square	3.4
406 deep square	4.25
288 square	6.4
288 deep square	9.0
242 square	9.0
200 round	9.0
200 square	11.0

*Presented by Shawn Laffe, Ball Seed Co., during the International Plug Conference held in Ames, Iowa, in November.

Delays in finishing are multiplied when 800 trays are held past their optimum transplanting time. Petunias are especially sensitive to holding, says Ron Derrig, Natural Beauty, Apopka, Florida. "Delaying transplant decreases the finished quality and increases growing time needed to finish the pack. This is extremely critical on petunias. As competition for light from neighboring plants increases, flowering is delayed because inadequate light is reaching the terminal bud," he says.

Cost to produce 800s is much less at 300 plants per square foot. It's so much less that 800s have the potential to alter the economics of bought-in plugs for flat producers.

Exactly where do 800s fit in?

For growers growing their own plugs, 800s offer an economic way to fit more trays on the plug bench during peak sowing season. "In early spring a grower may start in 400s or 512s. When space gets short they may go into 648s or 800s to get better space use in midseason and switch to a 128 or 200 at the end of the season when they want a larger plant that will finish fast," explains Shawn Laffe, Ball Seed Co., West Chicago, Illinois.

The biggest bonus in 800s comes for growers buying in plugs. "With 800s you can economically go to a 72 flat with a quicker finishing time than 648s," says Dave Tagawa, Tagawa's Greenhouses, Brighton, Colorado.

But because they cannot be held, success for 800s as a bought-in plug will hinge on planning: precision in planning production space use and transplanting line workloads.

Dave Sasuga, The Plug Connection, Vista, California, says 800s will not be produced for sale to other growers on speculation. "Eight hundreds will be grown to order only. You can't hold them. One of the effects they'll have is on production planning, and that will be good," he says.

Ball Seed's Gary Falkenstein sees 800s as a good size for growers buying in plugs for their first bedding turns. "With an 800 you can go in and provide a full range of plants for the first turn of bedding. With later crops you supply a bigger plug for faster finish."

Growers taking a look at buying in 800s will have to weigh the advantages of their lower price points with their longer finishing times in the pack. "Since 800 plugs are younger than 400 plugs, growers must decide if the decrease in price offsets the additional growing time and expense needed to finish the pack," Ron Derrig says.

It's a question for each grower

The 800 plug will get the attention of lots of growers. For some, it offers an economic first turn as a bought in plug, then perhaps they choose to use their own plug benches to supply larger plugs for subsequent turns. For others, it's a valuable size during peak production when sowing schedules and plug benches are jammed.

For all growers working with 800s, the real challenge is in scheduling finishing space and transplanting, because when 800s are ready, they don't wait.

99

Debbie Hamrick is editor of GrowerTalks *magazine.*

Gearing up for turn-of-the-century plug production

December 1989

by Julie A. Martens

VPD irrigation, sunrise shock treatments, water quality tests—these were the headliners at the 1989 International Plug Conference. Here's your guide to the buzzwords for plug production in the 90s.

Approximately 88 percent of total bedding plant production is from plugs. In the last 10 years, the number of plugs used in floriculture have increased from 500,000 in 1978 to roughly 3.5 billion in 1988. As a gambler, you might call plugs a "sure thing." They're definitely not the longshot in the industry.

The nearly 400 growers, researchers and allied tradespeople who gathered in Ames, Iowa, for the International Plug Conference in early November all see plugs as a key to the future—a key to increased crop turns, increased customer satisfaction and increased profits. Now it's your turn to stand on the cutting edge of plug technology.

Plug conference highlights

• **Water quality** is so variable between greenhouses that it's the overriding factor in making fertilizer decisions. You cannot grow plugs without knowing your water quality—test it at least every six months. According to Allen Hammer, Purdue University, there's no way you can make intelligent decisions about nutrition problems without knowing your water quality.

What should you shoot for in your water? Bicarbonates (alkalinity) around 61 ppm, soluble salts less than 1 $mmho/cm^3$, pH 5 to 5.5, calcium at 50 ppm, magnesium around 25 ppm, chloride less than 20 ppm, sodium under 40 ppm and boron 0.05 to 0.1 ppm.

• **The trick to DIF** is those sunrise shock treatments—when you pull your energy curtains abruptly open at dawn and let those cool, early morning temperatures work for you on plug height control. In addition to opening those curtains in the early a.m., you also need to keep the day temperature down to get the greatest height response, according to Royal Heins, Michigan State University. Don't be afraid to open those curtains, he urges. A little blast of cold air at sunrise could be your best offense in controlling plug height.

• **Disease pressure and cell number**—is there a relationship? Concern about the 800 plug tray includes questions about disease spread. Although there is no one clear relationship, one fact is known: The compartmentalization in a plug tray helps prevent disease spread. So, your arsenal to preventing disease doesn't include plug cell size. The most important thing to remember about diseases and plugs is that, with increased stress, susceptibility to diseases increases.

Make sanitation and optimum environments the rule for avoiding disease. "As long as we don't control dust, thielaviopsis will be perennial in our greenhouses," says Chuck Powell, Ohio State University. Old plug trays and plant debris must also be eliminated to beat black root rot. If you suspect thielaviopsis, tell your pathology lab when you send in samples. Otherwise, they won't always do the microscopic root exam needed to confirm diagnosis.

● **Weekly media tests** are crucial to your plug growing success. Check your media pH and soluble salts weekly; learn how production practices affect your media and these measurements. Good records will help sort out any problems you may run into in the course of a season.

● **In filling your plug trays**, you want to aim for uniform media levels in each cell, a "just right" moisture content and limited compaction. Strive to get what Shawn Laffe, Ball Seed Co., calls a "pillow effect" in each cell. You want the soil surface to be concave so that it catches the seed. How do you get that pillow? Try adjusting your brush height or using a dibbler. Just be careful not to compact the soil too much.

● **Wondering where Bonzi fits** into your plug production? If you want to use the new chemicals like Bonzi and Sumagic, conduct trials, keep records and have a few untreated plants you can use to make comparisons. Your best bet with these chemicals, says Jim Barrett, University of Florida, is to use low rates and multiple applications. Work your way up to a single application.

● **VPD irrigation**—watering crops based on vapor pressure deficit—is **the** way to water plugs, according to Tom Smith, Four Star Greenhouses. Sensors in the greenhouse measure relative humidity, air temperature and solar gain and transmit these measurements to a computer that alculates VPD and triggers a boom to irrigate. It's watering based on evapotranspiration rates—watering based on plant water use. The technology for VPD irrigation will soon be prevalent; this is definitely hands-off watering.

● **Every seed counts** when sowing plugs. "The bedding plant crop is one of the most lucrative crops we can grow," says Mel Sawaya, Fernlea Greenhouses, Ontario, "and plugs give it a boost. Take advantage of the plug bonanza. Plugs have been proven, time after time, to increase bedding plant profits. Maybe it's time you plugged into those profits, too."

"

Julie A. Martens is managing editor of GrowerTalks *magazine.*

Plugs: problems, concerns and recommendations for the grower

December 1989

by Dave Koranski and Paul Karlovich

If you're a plug grower, you know of the problems and concerns experienced with the major crops. Here are recommendations from Iowa State to solve major problems in growing quality plugs.

Plug production continues to roll along as a major growing system for bedding plants. Most growers now use the plug method of growing as the old, broadcast seed germination method is used less and less. Much information has been written on plugs and production in the past few years. Yet there is still much we do not fully understand. This article focuses on special problems and their solutions. The information presented is the result of many greenhouse visits across the country and from experiments conducted at Iowa State University.

Major crops and their major problems—here are solutions

Growers experience common problems on certain crops. The most frequently experienced difficulties involve diseases or disease-like symptoms as a result of culture-related stresses.

Vinca. Yellow, immature leaves occur when soil pH is greater than 6.8, a level that ties up iron. Acid may lower medium pH. Foliar applications of iron chelate can provide a temporary, quick source of iron.

Stem rot, caused by *Rhizoctonia*, infects seedlings at or just below the soil line and can be related to too much covering around the seed. Lodged seedlings are evidence of the disease. Warm, wet conditions can increase disease incidence. Control the disease by reducing the amount of covering using a fungicide such as Benlate, Banrot, Chipco 26019 or Terrachlor, and by practicing general sanitation.

Thielaviopsis fungi can be a problem shortly after transplanting. Symptoms of thielaviopsis include yellow bottom leaves, loss of root hairs, brown discoloration on roots, and slow plant growth or death. The root decline looks like pythium, but infected roots have no odor, and they have a water-soaked appearance. Disease incidence is increased by high pH, high salts and high temperature stress.

Achieve control by using a medium with a pH of 5.5 to 5.8 and an electrical conductivity less than 1.0. Leach thoroughly when watering. Use only new trays and media. Maintain a soil temperature of 62 to 78 degrees. Allow the media to dry slightly between waterings. Benlate and Cleary's 3336-F are cleared for use on this fungus. Do not use Subdue because it can enhance the development of this disease. Good sanitation practices in your greenhouse are effective deterrents to disease spread. Remove infested plants from the greenhouse. After transplanting, withhold additional fertilizer and fungicides until roots emerge from the root ball to prevent damage to sensitive root systems.

Pythium root rot infects root tips and then progresses upward along the root.

Table 1. Effect of pH on the incidence of *Thielaviopsis* on pansies.

Plant age	pH					
	5.0	5.5	6.0	6.5	7.0	7.5
2 weeks	0^z	0	2	5	4	2
4 weeks	0	1	4	7	8	6
6 weeks	1	2	4	8	6	7
6 weeks^y	0	3	3	3	2	0

^{z}Rating scale zero to 10 with 10 being the most affected plants.
^{y}Plants treated at 4 weeks with Banrot at 8 ounces per gallon.

Infected roots are soft and dark and have a strong musty odor. Best pythium development occurs under wet, cold conditions.

Control pythium root rot with similar cultural methods as described for Thielaviopsis control. Both Chipco 26019 and Truban can be used. Again, do not use Subdue.

Hold vinca at a soil temperature of 68 degrees F. Fertilize with a balanced formulation at 75 to 100 parts per million. Maintain proper pH and EC. Inspect plants for disease.

Marigolds and geraniums. pH levels less than 5.8 can cause excessive levels of some micronutrients, such as manganese, iron, sodium and zinc. Toxic levels of these nutrients are expressed as speckling on the foliage, necrosis or destruction of the growing tip.

Geraniums. Hold geraniums at 60 to 62 degrees F soil temperature. Prolonged or too low soil temperature may delay flowering, particularly if the plants have not initiated flowers. Fertilize with a balanced formulation at a moderate rate (150 to 200 ppm). Allow the soil to dry down, but do not allow the plants to wilt. Inspect plants for disease, such as Botrytis, yellow bottom leaves, and root discoloration. Botrytis is a problem during and after Stage IV. Good air circulation is beneficial, and Benlate, Daconil and Termil can be used for chemical control.

Pansies and petunias. Terminal bud abortion, side shoot proliferation, stunted growth, and hardened, distorted, cupped, mottled, and/or strapped leaves indicate boron deficiency. A soil pH above 6.0 can cause this problem by tying up boron. Boron deficiency has been misdiagnosed as Bonzi or thrips damage and occurs most often in hot, sunny weather when frequent watering can raise pH and leach boron. Boron deficiency can be corrected by lowering soil pH or by one application of soluble trace elements at one-half the recommended rate.

Control boron levels by maintaining a 5.5 to 5.8 pH and supplementing with boron in the fertilizer program (STEM at half the rate, one application is satisfactory). Under high temperature stress, STEM at full rate may be necessary. Solubor (0.25 ounce per 100 gallons) as a foliar spray can be used as a corrective measure.

Thielaviopsis is a problem on both pansies and petunias. Control is the same as for vinca. Table 1 shows the effect of pH on disease incidence on pansies. Low pH is an effective control measure for this disease. Pythium and rhizoctonia can also be a problem. Control is the same as for vinca.

In petunias, a high light level (400 footcandles) has been shown to reduce the amount of watery stem problems. With pansies, germination may benefit from light covering and with moderate moisture during Stage I. Moisture must be reduced when the root emerges to encourage growth and development.

Impatiens. Impatiens show a bewildering array of responses to light and dark. Some of the lots we tested showed no differences, small differences and large differences in response to light and dark (Table 2). In no case, however, was germination higher in the dark than in the light. In other tests conducted at two locations this past summer, there was no difference in light and dark germination tests on a number of cultivars.

Our experience shows us that commercial production practices **normally** do not show differences between light and dark germination. This is probably because dark conditions are never absolutely dark and the light requirement, if any, of impatiens is small. In some isolated cases, however, we have seen germination problems that do appear to be light related, and we feel lighting should be used as an insurance policy.

Solving plug tray problems

Plug trays vary in size and shape of the cells. The most common plug trays have from 128 to 800 cells. Plug cells may be round or square. The different sizes require adjustments in culture.

The key to successful growing in any plug tray is water management. Plugs must never be allowed to dry completely. Saturated media, however, contains almost no oxygen (zero to 2 percent), and therefore, cells must dry out to allow increased oxygen for optimum germination and growth. The smaller the cell size, the more vulnerable the plug is to fluctuations in moisture, nutrients, oxygen, pH and soluble salts in the soil. Deeper plug cells provide more drainage from the medium and better aeration.

800 Plug trays. The concern with this tray is the extremely small soil volume. Careful monitoring of the pH and soluble salt levels is important to ensure a good root system. Take care to avoid temperature, fertilizer, nutrient and water stresses, which can cause transplant establishment problems.

Scheduling is also different in 800 trays. One- to four-week longer cropping times, as compared to a 406 plug, are common with 800 plugs (Table 3).

Answers to seed quality

Many advancements have been made recently in seed quality. Seed can be modified to improve handling by the seeder: Marigolds are detailed or graphite-coated, tomatoes are de-fuzzed and begonias are pelleted.

Primed seed has been pre-conditioned to germinate. Under controlled condiions, the seed imbibes water and begins the first phase of germination. This process is stopped before the root penetrates the seed coat, and the seed is then dried back to its original moisture content. The result is the ability to germinate over a much broader range of temperatures. Germination is much faster, higher and more uniform than with refined seed.

Cultural practices for successful germination

The necessity for optimal levels of moisture, oxygen and temperature during germination cannot be stressed enough. Light and nutrition may also be beneficial for the germination of some crops, and certainly for growth later.

During the germination phase, including Stage I and early Stage II, the requirements of the seed for temperature, moisture and oxygen change. Initially in Stage I, the seeds on top of the soil need to be surrounded by high levels of humidity and oxygen. Once the root emerges (Stage II), the

Table 2. Effect of absolute darkness on impatiens germination.

Cultivar	Germination percentage Light	Dark
Rose Star	95	96
Super Elfin Lipstick	98	92
Accent Pink	97	80
Accent Salmon	89	61
Super Elfin Orchid	85	71
Super Elfin Orange	90	66
Accent Rose #1	94	64
Accent Rose #2	95	52
Impulse Rose	68	62
Super Elfin Red	86	42
Super Elfin Coral	98	25
Super Elfin Pink	96	26

Table 3. Weeks to finish from a 406 and 800 plug in a 48 flat.

Species	406 plug	800 plug
Ageratum	5-6	7-8
Alyssum	5-6	7-8
Begonia	6	8-9
Coleus	4-5	7-8
Impatiens	4-5	6-7
Marigold, African	5-6	6-7
Marigold, French	4-6	7
Pansy	5-7	8
Petunia	5-6	6-7
Snapdragon	5-7	6-7

requirements for oxygen increase. To supply the needed oxygen, the amount of moisture must decrease.

Temperature information can be found on seed packets or in seed catalogs. The temperatures, however, are only suggested and need to be tested and modified under individual germination conditions. For uniform, rapid germination and growth in Stages 1 and 2, each crop requires an optimum, uniform soil temperature. A recording thermometer, or thermograph, is helpful in monitoring air temperature. Soil temperature can be measured with soil probes or small soil thermometers.

Most seeds require temperatures between 70 and 80 degrees F to germinate. Seedlings usually remain in the chamber from two to 14 days, being removed when the seed leaves emerge. If left in the chamber too long, the very warm temperature, moist atmosphere and low light intensity cause stretching and non-uniform development.

Lights provide a source of heat and energy for photosynthesis. The benefits of lights become apparent during Stage 2 as the stem and seed leaves develop. Lights prevent excessive elongation and help promote branching.

Supplemental nutrition can enhance germination. During Stage I, an application of 25 ppm potassium nitrate may help to overcome seed dormancy in certain crops. A second application of potassium nitrate in Stage II, or 50 ppm of 20-10-20, provides nutrients beneficial for early seedling development. Petunias and begonias are two crops that respond well to early feeding.

Table 4. pH change over time in commercial media watered with water containing an alkalinity of 40 ppm.

		pH			
Media	Initial	3 days	1 week	2 weeks	3 weeks
A	5.5	6.5	7.2	7.5	8.1
B	5.7	5.8	6.1	6.2	6.5
C	5.6	5.6	5.8	6.1	6.0
D	6.4	6.2	6.3	6.5	6.4
E	5.1	4.9	5.1	5.3	5.4

Carefully consider medium

Growers should carefully consider the media they will use. A desirable medium should have a good balance between water-holding capacity and air porosity. Usually the higher the waterholding capacity the lower the air porosity and vice versa, so a good mix will balance these two properties.

Peat is the major influence on these properties. It should be fibrous, contain a range of particle sizes, and have a minimum amount of dust. The oxygen demand increases during Stage II. Aeration is less of a problem if the amount of moisture applied is decreased (use fog or fine mist to maintain high humidity without waterlogging the medium) or if the medium is allowed to dry as much as possible between waterings.

Table 5. Effect of bicarbonate concentration on the pH of growing media in different plug sizes.

	pH^z			
Bicarbonate level		plug tray size		
Meq/l (ppm)	200	406	512	800
2 (122)	5.7	6.2	6.1	7.5
4 (244)	5.9	7.1	6.9	7.8
6 (366)	5.9	7.5	7.6	8.1
8 (488)	6.0	7.5	8.1	8.5

^{z}All values are four weeks. Initial pH is 5.5.

The chemical properties of the medium are equally important. A starting pH around 5.5 to 5.8, soluble salts below 1.0 mmhos/cm^3 EC, and a high cation exchange capacity are best for plug media. Many commercial mixes contain a nutrient charge. A balance in nutrients is recommended at an approximate ratio of 1 nitrogen:1 potassium:1 calcium:0.5 magnesium.

Commercial media generally start at a reasonable pH, but great differences have been found in the way they react over time (Table 4). Some mixes maintain the pH very well, some rise gradually, and some show a pH rise as fast as 1 pH unit per three days. The test listed in Table 4 was done using water with 40 ppm bicarbonate, which shows that alkalinity did not cause the pH increase. The reason for these differences is probably due to differences in the amount of lime added.

Every batch of media is unique, and it cannot be assumed that one order of media is the same as the next. We strongly recommend that growers regularly monitor soil pH, so that adjustments can be made if necessary. Also, check the texture of each batch of media. Less fibrous mixes will have lower air porosity and watering schedules should be altered accordingly.

Symptoms seen with high pH and poor aeration are plants turning yellow, poor shoot and root growth, and a lack of root hairs.

Poor water quality can defeat your efforts

More and more attention is directed to water quality. Poor quality water can defeat the grower's efforts not only at germination, but also during growing on. Water should be tested by an independent laboratory. Desirable water qualities include:

● **pH 5.8 to 6.0.** This is the pH range at which most nutrients are soluble and available.

● **Soluble salts or EC** should be less than 1.0. If salt values are higher, determine which salts are present.

● **Bicarbonate or alkalinity** affects the pH of the growing medium. If the bicarbonate level exceeds 2meq/1g or 122 ppm, it can raise the pH. The smaller the plug cell, the more detrimental high bicarbonate levels (Table 5). Smaller cells are watered more frequently and leached more heavily, and thus pH tends to rise higher in these cells. Once the level of bicarbonate is determined, the amount of acid (sulfuric, phosphoric or nitric) needed to neutralize the bicarbonate can be calculated.

● **Sodium absorption ratio** (SAR) is the ratio of sodium to calcium and magnesium. If the SAR is less than 2 ppm and sodium is less than 50 ppm, the ratio should be adequate.

● **If chlorides exceed 30 ppm,** poor root and plant growth may result.

● **Boron levels greater than 0.5 ppm** have been shown to cause bud abortion in impatiens and petunias.

● **Minimum amounts of calcium** and magnesium must be available to balance the level of sodium.

Test results should be reviewed with a water quality expert. High bicarbonate or SAR levels may require implementation of reverse osmosis or deionized water systems. **"**

Dave Koranski is floriculture professor and Paul Karlovich is a recent Ph.D. graduate, Iowa State University, Ames.

How to grow 32 plug crops *by Dave Koranski*

Crop	Covering	Stage 1 Germ Temp* (°F)	Stage 1 Germ Moisture (% RH)	Stage 1 Germ Light	Stage 1 Time (Days)	Stage 1 Temp (°F)	Stage 2 Light Supp.	Stage 2 Moisture (% RH)	Stage 2 Fert. (ppm)	Stage 2 Time (Days)	Stage 2 Temp. (°F)	Stage 3 Light Supp.	Stage 3 Fert. (ppm)	Stage 3 Time (Days)	Stage 3 Temp. (°F)	Stage 4 Time (Days)	Stage 4 Temp. (°F)	Total Plug Crop Time (Weeks)	Comments
Ageratum		78-82	90-100		2-3	72-75		85-90	50, 1-2/wk	7	65-68		150, 1/wk	14	60-62	14	60-62	5-6	Root pruning necessary, may select early flowering cultivars.
Alyssum		78-82	90-100		2-3	72-75		85-90	50, 1/wk	7	65-68		100-150, 1/wk	21	60-62	7-10	60-62	5-6	Elongation of the seedlings may occur at temperatures above 75 degrees F. Preventative fungicide may be beneficial for soil-borne organisms.
Aster		70	85-90		4-5	68-70		85-90	50, 1/wk	7	65-68		150, 1/wk	14	60-62	7-10	60-62	4-5	None
Begonia—Fibrous		78-80	95-100	●	6-7	72-78		90-95	50-100, 1-2/wk	21	70-75		150, 2/wk	21	60-62	14	62-68	8-9	When seedlings are in Stage 2, they tend to lose vigor. Check soil for low and high fertility. Ammonium should be less than 20 ppm in soil.
Begonia—Tuberous		75-78	95-100		7-10	70-72	●	90-95	50-100, 1/wk	21	68	●	150, 1-2/wk	21	60-62	14	60-62	9-10	Night interruption of at least 50 footcandles is necessary to prevent tubers. Temperatures of 75 to 78 degrees F critical for germination. 70 to 72 degrees F for Stage 2. Ammonium fertilizer may inhibit growth.
Browallia		75	95-100		4-5	70-72	●	85-90	50, 1/wk	14	68-70		100, 1/wk	14	60-62	7	60-62	6-7	None
Cabbage	●	65-70	80-85		3-4	62-65		85-90	50-75, 1/wk	7	62-65		100, 1/wk	7	60-62	7-10	60-62	3-4	A soil medium with a lower nutrient charge can help prevent stem elongation.
Celosia		75	95-100		4-5	72-75		85-90	50, 1-2/wk	7	65-70		100-150, 2/wk	14	60-62	7-14	60-62	5-6	Salt levels and moisture levels should be maintained to avoid premature flowering. Use preventative fungicides. Stage 1 temperature is critical for uniform germination.
Coleus		72-75	90-95		4-5	72-75		85-90	50-75, 1-2/wk	10	68-72		150, 2/wk	14	60-62	7	60-62	5-6	Growth regulators and lower finishing salt levels help to keep plant compact.
Cyclamen	●	65-68	90		21-28	70		80-85	50-75, 1-2/wk	21	68-72		100-150, 1/wk	21	60-62	7-14	60-62	10-14	Germination medium pH should be higher than 5.5. Ammonium should be less than 10 ppm on soil. Oxygen is important for germination. Should be grown in 200 tray.
Dahlia	●	78-80	90-95		3-4	68-70		85-90	50, 1/wk	7	65-70		100, 1-2/wk	7	60-62	7	60-62	3-4	Difficult to sow mechanically. Low salt levels may be beneficial for growth and development.
Dianthus		70-75	95-100		3-5	70-75	●	85-90	50, 1/wk	7	65-70	●	150, 1/wk	14	60-62	7-14	60-62	5-6	Lower salt levels to prevent burn. Night interruption with 50 footcandles light can be used to promote flowering. Watch for stem elongation.
Dusty Miller		72-75	85-90		4-6	70-75		80-85	50, 1/wk	14	65-70		150, 1-2/wk	21	60-62	7-10	60-62	6-7	Low germination percentage; use multiple seed. Moisture levels on dry side after Stage 1 may be helpful. Preventative fungicide for alternaria is suggested.
Gazania	●	75-78	85-90		5-7	68-72	●	80-85	50, 1/wk	7	65-70	●	100, 1/wk	17	60-62	7-14	60-62	5-6	Should be grown in the larger plug tray.
Geranium	●	70-75	90-95		3-5	70-75		80-85	50-100, 1/wk	10	65-70		150, 2/wk	14	60-62	14-21	60-62	6-7	pH must be maintained at 5.8 or greater to prevent iron manganese, sodium and zinc toxicity. Germination temperatures higher than 75 degrees may be detrimental.
Gerbera	●	68-72	90-95		5-6	68-72		80-85	50-100, 1/wk	14	65-70		150, 1-2/wk	14	60-62	7-14	60-62	6-7	None

Crop	Covering	Stage 1 Germ Temp* (°F)	Stage 1 Germ Moisture (% RH)	Stage 1 Germ Light	Stage 2 Time (Days)	Stage 2 Temp (°F)	Stage 2 Moisture (% RH)	Stage 2 Light Supp	Stage 2 Fert (ppm)	Stage 2 Time (Days)	Stage 2 Temp (°F)	Stage 2 Light Supp	Stage 3 Fert (ppm)	Stage 3 Time (Days)	Stage 4 Temp (°F)	Stage 4 Time (Days)	Total Plug Crop Time (Weeks)	Comments
Impatiens		75-80	90-95	●	3-5	72-75	80-85	●	50-100 1/wk	10	68-72	●	100-150 1/wk	14	60-62	7-14	5-6	Keep Stage 1 wet for first two to three days, then decrease moisture. Use growth regulators to maintain compact growth if fertilizer is used. HID lights for two to three weeks will provide vigorous plant with thicker stem.
Lisianthus		75	85-90		10-12	68-72	80-85		50-75 1 (wk)	21	65-70		100 1/wk	21	60-62	7-14	9-10	None
Lobelia		75-80	95-100		4-6	68-72	80-90		50 1/wk	7	65-68		100 1/wk	14	60-62	7-10	5-6	Good air movement and fungicides can be used to prevent damping off and other fungal diseases.
Marigold—African	●	75-80	90-95		2-3	68-70	80-85		50-75 1/wk	5	62-65		100-150 1/wk	14	60-62	7-10	4-5	Need two to three weeks to flower. pH levels below 5.8 result in toxic levels of iron, zinc, manganese and sodium; appears as necrotic spots and burned edges on lower level.
Marigold—French	●	75-80	90-95		2-3	68-70	80-85		50-75 1/wk	7	62-65		100-150 1/wk	14	60-62	7-14	5-6	Phytotoxicity can appear from iron and manganese deficiency.
Pansy	●	62-68	95-100 / 90-95		4-7	62-68	75-80		50 1/wk	7	60-62		100 1/wk	14	55-60	14-21	6-7	Stage 1 65 degrees F and high moisture for three days, then reduce moisture. Medium with low nutrient charge and low phosphorus helps to prevent stem elongation.
Pepper	●	70-75	90-95		5-7	70-75	80-85		None	7-10	65-70		100 1/wk	10	60-62	7	4-5	Sensitive to damping off if plants kept too wet or cold during Stage 1 and 2.
Petunia		75-78	90-95	●	3-5	72-78	75-80	●	50-75 1-2/wk	14	62-68	●	150 1-2/wk	14	62-65	7-10	5-6	Near saturation moisture level critical for first two days of germination, then decrease moisture thereafter. Medium pH should be less than 6.8 to prevent iron deficiency.
Portulaca		75-80	95-100		2-3	70-72	85-90		None	7	60-65		100 1/wk	14	60-62	7-14	4-5	Use growth regulators early to avoid elongation. Use preventative fungicide for damping off disease.
Primula	●	62-68	95-100 / 90-95	●	7-10	60-65	85-90		25-50 1/wk	14	60-62		100 1/wk	35	60-62	7-14	9-10	None
Salvia	●	75-78	90-95	●	5-7	70	80-85		25 1/wk	7	62-65		100 1/wk	21	60-62	7	5-6	Refined seed may improve germination. Low fertilizer levels during Stage 1 and 2, increase in Stage 4. Do not hold plugs.
Snapdragon		70-75	90-95	●	5-8	65-70	80-85		None	14	62-65		100 1/wk	14	60-62	7-10	6-7	Alternating 70 degree day and 80 degree F night temperatures may be beneficial for germination. Use low salt levels.
Tomato	●	70-72	95-100		2	68-72	85-90		None	7	60-65		100 1/wk	7	60-62	7	3-4	May stretch when germinated without lights. Moisture stress is effective to control height.
Verbena	●	75-80	95-100 / 2 days / 75-85		4-6	72-75	75-80		25 1/wk	14	68-72		100 1/wk	14	65-68	7	5-6	Dry conditions during germination may be beneficial. A preventative fungicide program may improve germination.
Vinca	●	75-80	90-95 / 3 days / 75-85	●	4-6	72-78	75-80	●	25 1/wk	14	68-72	●	100 1/wk	21	65-68	7	6-7	Cover seed, but sparingly. Maintain temperature high moisture for three days, then reduce.
Zinnia	●	80	95-100		2-3	70-72	80-85		None	7	65	●	100 1/wk	7	60-62	7	3-4	Use a preventative fungicide for alternaria.

*Degrees C equals 5/9 (F minus 32).

Dave Koranski is professor of floriculture, Iowa State University.

Special comments:

Covering: Certain seed need to be covered to exclude light. In these seed light causes desiccation, so they do not maintain adequate moisture levels to germinate. Germination percentages decrease significantly under lighted conditions.

Fertilizer: Fertilizer is in parts per million (ppm) of nitrogen and it is coming from a commercial 20-10-20 fertilizer with calcium added. Fertilizer also may come from adding separate components NH_4NO_3, $Ca(NO_3)_2$ and KNO_3. The components need separate injector heads to prevent precipitation.

If ammonium is used in Stage 2, be careful to keep temperatures above 68 degrees F.

Stages 3 and 4 fertilizer may be 20-10-20 as long as temperatures are above 68 degrees F. Because most 20-10-20 fertilizers lack calcium, it may be necessary to use a $CaNO_3$ fertilizer in addition but separately.

When a toned plant is desired, $CaNO_3$ and KNO_3 may be used

Definition of Stages

Stage 1 — time of radicle emergence.
Stage 2 — stem and cotyledon emergence.
Stage 3 — growth and development of true leaves.
Stage 4 — plants ready for transplanting, shipping or bedding.

Explanation Moisture Levels Stage 1 and 2

The percent relative humidity moisture levels listed for Stages 1 and 2 are approximate. Initially in Stage 1, the seed needs to be surrounded by high levels of humidity and oxygen. Once the radicle emerges, Stage 2, and requirement for oxygen increases the amount of moisture must decrease.

Take a look at costs before you decide to grow your own or buy in

December 1988

by Cindy Quincy

Time and time again plug users wonder if they should grow their own plugs or buy them from a plug specialist. Most growers will agree that using plugs in their bedding plant and perennial production is the way to go. Growers recognize the benefits of increased production per square foot, easier and quicker transplanting with less skilled labor and overall improved quality and uniformity of product in the final container. All are a real plus for the plug user, giving savings in time and money.

But what if you grow your own? Do you enjoy the same savings in time, costs and labor? Some growers insist that plugs are too expensive to buy. When you ask those same growers how much it would cost them to produce their own plugs, they don't know, but somehow feel certain it would be better and cheaper than buying plugs. Or is it?

Costs vary from grower to grower

Before you decide whether you should grow your own plugs or buy them from a specialist, be aware of the costs involved—both the direct and indirect costs.

Based on conversations with a range of plug producers, one thing is clear to me: costs vary widely from grower to grower. Because of this variation, I will not attempt to give you exact cost figures.

The cost to buy in an impatiens plug ranges from 3 to 7 cents delivered. How is this? What are the variables—plug quality, method of costing, conditions of sale?

Plug specialists offer a range of choice and services, from bread and butter items with high minimums to a feast of choices with low minimums. The market each plug specialist targets affects their costs, assuming they know what their costs are.

From my own experience with plug production, I would say to anyone considering growing their own plugs, before you begin to absorb yourself in the time-consuming effort of costing, ask yourself several questions:

- Do you have a highly skilled grower?
- Are you prepared to take this highly skilled grower and devote his attention to plug production? You'll also need a highly skilled sower. Plug tray quality is often made or shattered at the seeder.

Once you've committed to a skilled grower and sower, take a look at the equipment you need. First, you'll want to invest in a seeder, a flat filler and a plug dislodger. Then there's the investment in redesigning your facilities to accommodate plug production. Several things to consider are:

- Do you want to go with a germinating chamber with a fogging system and lights?
- What about a bench system with bottom heat, shade curtains, boom irrigation and HID lights?
- What about the benefits of heated water?
- Are you in an area that requires reverse osmosis systems to remove salts and to lower the pH of irrigation water?

Covering the basics: A cost analysis of growing plugs

The area in greenhouses for growing plugs will be among the most capitalized square feet you own. At no other point in the crop's life are plants so densely spaced or does each square foot represent so much in crop value.

For example, if it costs you 3 cents to produce one plug, a common 512 plug tray represents $15.36. It pays to make the most use of your plug space by investing in tools to do it right.

You'll need carts for staging plug trays in the germination chamber and for transporting from Phase I to Phases II and III unless, of course, you can afford mobile benching.

For spring bedding crops, you'll often need to start plugs as early as November or December. The extra fuel and electrical costs to maintain 75- to 85-degree F temperatures for germination will figure in your overhead. You need to consider what crops you are displacing to make room for plugs and if it would be more profitable to grow those crops and buy in plugs.

Next, look at your cost of materials. For the sake of time and for some meaningful figures I suggest you divide your costs between fixed and variable, with seed being a distinct variable cost that you consider each time you figure plug production costs.

Though soil costs can vary from a 406 to a 128 tray, the impact it has on a per plug basis is not great enough to consider it a variable each time you cost out a crop. Fixed costs for each plug crop apply to costs of the tray, fertilizer, water, label and vermiculite.

Determining seed cost is not so simple as looking up the price in your seed catalog. The type of seed you use, such as standard, high energy, pelleted or pregermed, will affect the cost.

Germination rates and built-in oversows or loss factors must be considered, along with plug size and the decision to single or multisow. Germination rate must not be dictated by what is printed on the seed packet; it has to be determined by what your grower and facilities can generate.

Final production finances include labor and overhead

Plugs are a hi-tech crop and require hi-tech seed, people and facilities. Having only one of these factors will not eliminate the need for the other two.

Once you have determined the cost of materials, add an overhead production figure of approximately 20 percent. Determine the cost of labor, which is the most difficult, but which is figured as a fixed cost. Consideration is given to sales/clerical, pregrowth and growth stages. An overhead of approximately 50 percent is added to this total.

Sum your variable and fixed costs and divide this figure by the number of saleable or transplantable plugs to find your cost per plug.

The specialist plug grower has added costs that the in-house producer doesn't: shipping, marketing, advertising and commissions to brokers. Large plug producers, however, usually have the advantage of economies of scale to offset the additional costs, whereas the smaller grower does not.

Many growers choose to grow some plug crops and buy in more difficult or small-run items. Only you can decide if your needs, resources, facilities, technical skills and labor supply can sustain in-house plug production.

"

Cindy Quincy formerly owned and operated Green America, Blue Point, Long Island, New York.

Checking out plugs up close . . .

December 1988

by David S. Koranski and Shawn Laffe

Mechanization used to be the watchword in plug production, with the emphasis on efficiency in using time, space and labor. Keeping pace with current technology, growers are realizing benefits of improved crop uniformity and reduced shrinkage.

Every aspect of production has seen advances, from seeders to germination chambers to water analysis to growth regulators. The successful plug user may grow his own or buy in, may finish in cell packs or 4-inch pots, or may retail at home or ship 300 miles. To produce a quality crop, the plug grower must be aware of what's new in the basics.

Automatic seeders

A number of mechanical seeders are available, ranging from simple to complex in operation. The different seeders operate using one of several systems, including vacuum, electric, compressed air, water placement or rotary drum. Costs vary from less than $1,000 to over $30,000, depending on seeder complexity. The type of seeder to purchase depends on the grower's needs in volume and production methods and also upon his operation and financial investment ability.

For example, a beginning plug grower may initially choose a less expensive and less complicated seeder. As the beginning grower's economics and skills improve and needs change, a different seeder can be purchased.

To choose a seeder, growers should examine features that different machines provide and ask key questions.

- What types of plug trays will the seeder handle?
- Is the speed of sowing (trays per hour) variable or fixed?
- What types of seed will the seeder sow? Does the seed need to be detailed, de-fuzzed or pelleted?
- How much seed does the seeder hold? How much seed is required to operate the seeder?
- What is the ease of changeability of templates or manifold heads in order to sow different types and sizes of seed or trays?
- How accurate is seed placement? What percentage of empty cells can be expected?
- How accurate is the seeder in double-seeding cells?

Plug trays

Plug trays vary in size and shape of the cells. The most common plug trays have from 128 to 800 cells. Plug cells may be round or square. The different sizes of plug cells require adjustments in culture.

The smaller the cell size, the more vulnerable the plug is to fluctuations in moisture, nutrients, oxygen, pH and soluble salts in the soil. The deeper the plug cell, the more drainage in the medium, which allows better leaching of nutrients, reduced salt build-up and greater aeration.

The key to successful growing in any plug tray is water management. Plugs must never be allowed to dry completely, but moisture must be regulated to allow thorough wetting of the cell, drainage and oxygen exchange.

When saturated, the medium contains almost no oxygen (zero to 2 percent).

Mechanical seeders are available which use several operational systems, from vacuum to electric to compressed air to water placement to rotary drum. Growers purchase seeders based on their needs in volume and production methods, as well as on their financial investment abilities.

Therefore, cells must dry out to allow increased oxygen for optimum germination and growth. When the medium drips water only when squeezed, this indicates the right combination of moisture and oxygen for plant growth.

The type of plug tray to purchase depends on the use of the plug.

• What is the final container for the plant? Smaller plugs finish well in flats. Larger plugs produce attractive pots, baskets and planters.

• What about quality? Larger plugs produce a higher quality finished plant in a shorter period of time than smaller plugs do.

• What cell size best fits the crop? Succulent crops (begonias, impatiens) or crops that require a longer period of growth before transplanting (perennials, cut flowers, flowering pot crops) grow better in the larger cell sizes. Fast crops do well in the smaller plug cells.

• How can space be scheduled? Smaller plug cells pack a higher density in a given area (more plants per square foot) during peak production times when space is limited. Larger cells require more space per plant, result in a larger seedling, and are a logical choice for light production times.

• How can time be scheduled? Larger plug cells are more tolerant to stress than smaller plug cells, allowing the plants to finish faster.

• Square cells hold more soil than round cells, providing more capacity for root development. Additionally, water is distributed more evenly in a square cell than a round cell.

• What about shipping? Smaller plugs (512 or smaller) are not so tolerant of shipping due to small soil volume and need to be transplanted immediately.

Seed

Many advancements have been made in seed quality. Seed can be modified to improve handling by the seeder: marigolds are detailed or graphite-coated, tomatoes are defuzzed and begonias are pelleted.

Primed seed has been preconditioned to germinate. Under controlled conditions, the seed imbibes water and begins the first phase of germination. This process is stopped before the root penetrates the seed coat, and the seed is then dried back to its original moisture content. The result is the ability to germinate over a much broader range of temperatures. Germination is much faster, higher and more uniform than with refined seed.

Germination

During growth and development, the plug seedling goes through several stages. The stages have been identified as Stages 1, 2, 3 and 4.

• In Stage I roots emerge from the seed coat. This period of growth requires high levels of moisture and oxygen around the seed.

• After the root has emerged, the seedling enters Stage II, during which the root penetrates the soil, and the stem and seed leaves emerge. At this time, the amount of oxygen needed by the root increases, so the amount of moisture applied should be decreased.

• During Stage III the true leaves grow and develop.

• In Stage IV seedlings are ready for shipping, transplanting or holding.

The most critical time period for plugs is during Stages 1 and 2. The supply of optimal levels of moisture, oxygen, temperature and light is critical and will mean success or failure for the grower.

Germination facilities

Germination chambers: Germination chambers are specially designed rooms that provide a controlled environment for seed germination. The chamber walls can be constructed of exterior plywood and insulation. Insulation provides a buffer to outside temperature conditions, although fluctuations in outside temperatures can cause condensation inside the chamber.

The room can be divided into different environmental zones or compartments. Seed trays are typically placed in carts that can be rolled in and out of the chamber.

A mist or fog system can be installed to provide moisture and still maintain adequate oxygen levels for germination. The moisture droplet size should be 15 to 80 microns to maintain approximately 95 to 100 percent relative humidity with oxygen around the seed. Seed trays should not be hand watered because of the high potential of washing seed out of the cells or burying the seed in the media.

Cool white fluorescent tubes provide both light and heat to the plugs. Lights may be placed horizontally or vertically, although condensation on the lights is less of a problem when they are placed vertically. Lights should provide 200 to 400 footcandles at tray level.

A uniform temperature can be controlled by forcing conditioned air through a perforated false ceiling into the chamber. It is very important to limit air movement to prevent uneven distribution of the moisture. Temperature can also be maintained by placing hot water pipes around the outside walls of the chamber. The heated water regulates the chamber temperature and is also used in the pressurized mist system within the chamber.

From the plug tray to the finished product: Tips for transplanting

The environment in which the transplanted seedlings are placed is important. Healthy, actively growing plugs need to be placed in a medium that provides 15 to 20 percent aeration with soluble salts under 1.5 to encourage fast rooting-out. Containers can be prepared by filling with moistened media and predibbling holes for plugs.

Separating multiple seedlings may cause damage to seedling roots and is not recommended. Monitor water quality to ensure sodium absorption ratio levels are less than two to prevent destroying soil structure, reducing oxygen penetration and inhibiting roots. Fungicides can be applied on non-susceptible crops as early as one to three days after transplanting.

Some crops need individual attention at transplanting. Vinca needs a soil with no nutrient charge and 5.5 pH. Only after the vinca plug roots into the medium should a fungicide and fertilizer be applied to the plants.

Temperature. To finish the bedding flats, temperature should be set at 65 degrees F for the first four to five days until the plants are established. Once rooted, the temperature can be lowered to 55 to 60 degrees F to hold the plants or raised to 70 to 75 degrees F to force flowering. Temperature studies by Michigan State University can be applied to bedding plants. If the plants are too tall, height may be checked by maintaining warm nights and cool days.

Nutrition. Fertilizer can be applied with each irrigation or at intervals, using a high nitrate-nitrogen source of fertilizer at 250 to 300 ppm nitrogen.

A variation of the germination chamber is the sweat chamber, which does not use lights. Special care must be made to remove seedlings from a sweat chamber into light before the stem elongates. Also, because a greenhouse environment will be much brighter and drier than the sweat chamber, the seedlings should be transferred to a conditioning area before being moved into the greenhouse.

Greenhouse benches: Whether the seed trays are placed in a chamber or on a greenhouse bench, a controlled environment is necessary. If greenhouse space is available and if the grower can accurately control the environment, bench germination will be possible. Greenhouse benches are also adequate for the germination of crops which are not as sensitive to environmental conditions, such as zinnia, marigold or melons.

Benches should be level and constructed of expanded metal to allow air movement. This is especially important if under-bench heating is used. Either under- or above-bench heating should maintain proper soil temperatures. If root-zone heat is used, moist capillary mats over the heating cables may be needed to distribute the heat and prevent hot or cold spots. Roots can grow into the mats.

Moisture can be supplied by fog or mist systems that provide a very fine water droplet size. Boom watering systems in germination are recommended for small seed only if the droplet size is small (50 to 80 microns), producing a fine mist. A water droplet size of 15 to 80 microns is needed to allow adequate oxygen to reach the seed and newly emerged root. Propagation mist has a droplet size of 300 to 800 microns, which is too heavy to allow oxygen to reach the seed.

A very successful method for germinating on greenhouse benches involves covering the trays with porous non-oil-based material such as AgriCloth. This material provides an ideal microclimate for seed germination by not allowing moisture droplets to drown the seed. Coverings should be gently removed after germination once seed leaves unfold.

Cultural practices for successful germination

The necessity for optimal levels of moisture, oxygen and temperature during germination cannot be stressed enough. Light and nutrition may also be beneficial for the germination of some crops, and certainly for growth after germination.

During the germination phase, including Stage I and early Stage II, the requirements of the seed for temperature, moisture and oxygen change. Initially in Stage I, the seed on top of the soil needs to be surrounded by high levels of humidity and oxygen. Once the root emerges (Stage II), the requirement for oxygen increases. To supply the needed oxygen, the amount of moisture must decrease.

For uniform, rapid germination and growth in Stages 1 and 2, each crop requires an optimum, uniform soil temperature. Temperature information can be found on seed packets or in seed catalogs. The temperatures, however, are only suggested and need to be tested and modified under individual germination conditions. A recording thermometer, or thermograph, is helpful in monitoring air temperature. Soil temperature can be measured with soil probes or small soil thermometers.

Most seed require temperatures between 70 and 80 degrees F to germinate. Seedlings usually remain in the chamber from two to 14 days, being removed when the seed leaves emerge. If left in the chamber too long, the very warm temperatures, moist atmosphere and low light intensity cause stretching and non-uniform development.

Lights provide a source of heat and energy for photosynthesis. The benefits of lights become apparent during Stage II as the stem and seed leaves develop. Lights prevent excessive elongation and help promote branching.

Supplemental nutrition can enhance germination. During Stage I, an application of 25 parts per million (ppm) potassium nitrate may help to overcome seed dormancy in certain crops. A second application of potassium nitrate in Stage II or 50 ppm of 20-10-20 provides nutrients beneficial for early seedling development. Petunias and begonias are two crops that respond well to early feeding.

Germination medium

A growing medium must provide a suitable environment from the time of germination through the time of transplanting. During this time the plant changes in its size, shape and needs. The medium must, therefore, be flexible to meet the plant's needs, yet consistent in its physical and chemical properties. Plug cells contain a very small amount of medium that can fluctuate rapidly in moisture content, aeration, pH, soluble salts and nutrient levels.

A desirable medium should have a high water-holding capacity, a broad particle-size distribution, and high buffer capacity to resist chemical and physical changes. At one time, different media were recommended for different crops. Today, studies have shown that the same media can be used for different crops by practicing proper moisture management.

Characteristics important for a medium include:
● The ability to hold moisture for germination in Stage I, especially for pansy, vinca and salvia.
● A porosity level to provide adequate oxygen for a developing seedling root. The level of oxygen needed increases during Stage II. Aeration can be provided by decreasing the amount of moisture applied. Aeration of the original medium

Handling purchased plugs

● Open and unpack boxes immediately upon arrival. Carefully check the condition of the plugs. Check the packing list to verify that you have received what you ordered. Report any damage or discrepancies to your supplier immediately.
● Place plug trays on benches under light shade and water thoroughly with clear, tempered water, making certain that the plugs near the edges of the trays are well-watered. Be sure there is good drainage away from the bottom of the plug trays.
● Allow plugs to acclimate for 24 to 48 hours before transplanting.
● Maintain a minimum night temperature of 5 degrees F.
● After the initial watering, apply a general-purpose fertilizer (20-10-20) at 50 to 80 ppm every other watering.
● Transplanting and finishing procedures for purchased plugs are the same as for home-grown plugs.

can also be improved by adding calcined clay, perlite, or bark. A balance must be achieved between the water-holding capacity of the soil and aeration. As the water-holding capacity increases, soil porosity decreases.
- An electrical conductivity (EC) reading of 1.0 or less.
- pH of 5.8 to 6.5 to allow availability of both major and minor elements. Most commercial mixes contain a nutrient charge. A balance in nutrients is recommended at an approximate ratio of 1 nitrogen: 1 potassium: 1 calcium: 0.5 magnesium.

The medium should be tested at a soil laboratory. The key elements to monitor include (based on Iowa State University paste extract tests):

ammonium	< 20 ppm
nitrates	40 to 80 ppm
phosphorus	5 to 15 ppm
potassium	35-80 ppm
calcium	50 to 100 ppm
magnesium	25 to 50 ppm
sodium	< 50 ppm
chlorides	< 30 ppm
sulfates	< 100 ppm

Water quality

More and more attention is directed to water quality. Poor quality water can defeat the grower's efforts not only at germination, but also during growing on. Water should be tested by an independent laboratory. Several desirable water qualities include:
- pH 5.8 to 6.0 is optimum, as this is the pH at which most nutrients are soluble and available.
- Soluble salts or EC should be less than 0.75. If salt values are higher, determine which salts are present.
- Bicarbonate or alkalinity affects the pH of the growing medium. If the bicarbonate level exceeds 2 $^{meq}/_1$ or 125 ppm, it can raise the pH. Once the level of bicarbonate is determined, the amount of acid (sulfuric, phosphoric or nitric) needed to neutralize the bicarbonate can be calculated.
- Sodium absorption ratio (SAR) is the ratio of sodium to calcium and magnesium. If the SAR is less than 2 ppm and sodium is less than 50 ppm, the ratio should be adequate.
- If chlorides exceed 30 ppm, poor root and plant growth may result.
- Boron levels greater than 0.5 ppm have been shown to cause bud abortion in impatiens and petunias.
- Minimum amounts of calcium and magnesium must be available to balance the level of sodium.

Test results should be reviewed with a water quality expert. High bicarbonate or SAR levels may require implementation of reverse osmosis (R/O) or deionized water systems.

Growing on

The growing-on period for plugs, Stages 3 and 4, includes the time from development of the first true leaves through the time plugs are transplanted. This is when plugs are more flexible in environmental requirements, but conditions need to be uniform and controlled.

Temperature

Temperatures in Stage III are typically 60 to 65 degrees F, depending on the type of growth desired. If the plugs are to be shipped, they should be slightly toned or hardened to withstand shipping using a temperature of 60 degrees F. If the plugs are for in-house use, 65 degrees F will produce a satisfactory plant. During Stage IV, seedlings may be forced using temperatures up to 70 degrees

F or held back for up to two weeks by reducing temperatures to no lower than 58 degrees F. This technique works well with petunias. Growers should note that unless plants have already initiated flowers, temperatures below 60 degrees F will cause flower delay, as will too high temperatures.

Moisture
In Stages 3 and 4, water seedlings as needed, preferably with overhead mist. Apply enough water to cover the trays uniformly and to leach out soluble salts. The trays must also be allowed to dry sufficiently to allow oxygen into the soil and to let harmful gases, such as ethylene, escape.

Light
For certain crops, seedling growth and development benefit from at least 400 footcandles at plant level for 18 to 20 hours a day, two to three weeks after germination. High intensity discharge lights are recommended.

Long-day plants, such as dianthus and tuberous-rooted begonia, require interrupted night lighting of approximately four hours to flower on time. Short-day plants, such as African marigolds, require shading during this time to induce flowering.

Petunias need temperatures above 59 degrees F and daylength longer than 12 hours to induce flowering.

Growth regulators
Plant height can be controlled by cultural management or by chemical growth regulators. When used properly, chemical growth regulators control plant growth without causing flower delay. There are several points growers should follow when using growth regulators.
- Read and follow label directions.

Growing versus buying in: A key decision

Plugs have become a very marketable commodity. Growers can grow their own plugs, purchase plugs from another producer or supplement home-grown plugs with purchased plugs.

The availability of purchased plugs has eased some pressure that non-plug growers face to convert their operations to plugs to remain competitive. Plug production requires a strong commitment by the grower.

The initial investment in plug production demands capital for equipment and facilities. Though top-of the-line equipment and structures are not necessary the benefits of plug production will not be realized if the seeder, the water system, the lights, etc. do not function properly or if the environment cannot be controlled.

Likewise, a strong commitment must be made by the grower for quality. The type of product the grower needs may vary, but what grower does not need quality? The attention to detail and the willingness to learn and train as technology advances will enable a plug grower to be successful.

For growers who cannot afford the time and effort in growing their own plugs, they may choose to purchase plugs from specialized plug growers. The specialized producer has already committed the investment and time in plug production. Some growers may grow a limited variety of crops in a restricted environment; others may be able to produce a wide variety of quality crops by providing a range of environmental conditions.

Many growers opt to supplement their production figures with purchased plugs for many reasons. Tight production schedules can be opened up by purchasing plugs when space and time allow. Some growers excel in growing certain crops and do poorly in the production of others. To avoid possible losses in production and time, a less successful crop can be purchased.

Through buying in plugs, growers can take advantage of varied environments found in different regions of the country. If the weather is too hot for plug production in one region, plugs can be grown and shipped from a cooler region to a warmer one, expanding the marketing season for the finished product. A grower can also increase marketing flexibility by purchasing plugs to grow specialty varieties.

The bottom line of decision-making rests on the economics of the choices and the grower's ability to pay attention to detail and to understand plant growth and development. Quality plugs are represented by greater uniformity and less shrink. The grower must carefully examine the options. In the end, will it be more profitable to grow plugs or to purchase all or a portion of the plugs needed to fill out a product mix?

- Treat plants at the correct physiological stage. Typically, the first application is made when the growing point of the seedling begins to elongate and is about one-half inch in diameter or at the first to second true leaf stage.
- Plants should be actively growing and turgid when treated and not under any type of stress.
- Use correct concentrations required by specific crops and varieties.
- Apply the correct volume of chemical per given area, such as 1 gallon of solution spray per 200 square feet, or at a rate to deliver a certain concentration of active ingredient per container. It is not recommended to spray to runoff.
- More frequent applications at a lower concentration may produce more manageable control than one or two applications of a higher concentration. High concentrations may delay flowering or produce smaller flower size, especially when applied late in seedling development.
- The foliage should be dry at the time of application. Check the waiting period after treatment before water can be applied to the foliage. For Bonzi the wait period is one hour; for B-Nine, 24 hours.
- Note the environmental conditions, especially temperature and light. Temperatures of 65 to 70 degrees F and low light conditions are favorable for growth regulator applications.

Plant response to a chemical may vary with season, region and other environmental conditions. Records of calibrations, volume applied, weather patterns and plant response are critical in fine-tuning the use of chemicals. Specific chemical recommendations for a particular geographic region can be obtained from extension agents, manufacturers or trade journals. Conduct small-scale sample tests to determine the response under individual greenhouse conditions.

Nutrition

As plug seedlings grow and develop, nutrition management becomes more an art than a science. Crops have different requirements depending on species, cultivar, stage of development and scheduling.

A grower must take into consideration the cation exchange capacity (CEC) of the soil, type of greenhouse covering, water quality and amount of water applied with each irrigation.

- **Medium:** Different peat-based media have different CEC (ability to hold onto nutrients). Every medium should be tested before using. A grower will usually fertilize more in a medium with a lower CEC.
- **Crop:** Crops such as vinca and pansy are sensitive to fertilizer salts and should receive applications only as needed after roots are well established.
- **Stage of development:** Young Stage I and 2 seedlings can benefit from low (25 to 50 ppm) levels of fertilization once a week. Stages 3 and 4 involve active growth. Moderate levels of 50 to 100 ppm from potassium, ammonium and calcium nitrate with minor elements can be applied as needed, using caution on salt-sensitive crops.
- **Type of growth:** Lush plant growth can be achieved by using ammonium and potassium nitrate fertilizers, whereas calcium and potassium nitrate fertilizers promote toned growth.
- **Environmental conditions:** Warm temperatures that promote active and rapid plant growth will require more fertilizer applications to support plant growth. Low temperatures, such as those used to hold plant material, will require fertilizer application less often because the plants will not be actively growing. Plants actively growing under high light-transmittance coverings, such as acrylic or glass, will require more fertilizer than if growing under darker conditions.
- **Scheduling:** Fast-crop scheduling will require a higher level of fertilizer

than slow cropping or holding.

● **Moisture regime:** Heavy watering with a constant feed can suffice with a lower fertilizer concentration. Heavy leaching can wash away nutrients, including nitrates, phosphorus, calcium, magnesium and boron. Moisture and fertilizer applied less frequently will require a high fertilizer concentration with careful attention to thorough leaching of salts. Water containing high levels of salts may require some type of purification before fertilizer is added.

● **Soil testing:** Soluble salts and pH should be monitored weekly in both the soil and the water.

Finishing

From Stage IV, the plug seedlings are transplanted into the final container and finished for sale. Plugs are transplanted mechanically with special equipment or manually by crews of workers. Petunia and dusty miller lend themselves to mechanical transplanting.

Special plug dislodgers are recommended for manual transplanting to prevent damage to seedlings being pulled from the trays. Watering plugs two to three hours before transplanting helps in removing the plugs from the trays. After transplanting, seedlings should be watered in thoroughly.

"

Dr. David S. Koranski is professor of floriculture, Iowa State University, Ames, Iowa. Shawn Laffe works in Plant Research and Development at Ball Seed, West Chicago, Illinois. Special appreciation is extended to Dr. Roger Styer, Ball Seed, for his assistance in manuscript preparation.

Putting the plug boom into perspective

December 1988

by Debbie Hamrick

Plugs are here to stay

The plug unit, a small, pre-started, actively growing plant with a root system contained in a media to give roots form, is here to stay—at least for the forseeable future. Why?

● **Uniformity:** Greenhouse space is expensive. Using the plug system, whether you choose to grow your own or buy them in, allows more uniformly timed crops with more uniformity among plants in the finished flat. Patching and cherry picking not only takes time, but also requires labor.

● **Labor:** Plugs are a whole lot faster and easier to transplant than bareroot seedlings.

● **Time:** Time is money, and in the greenhouse, turns per square foot equal dollars in the pocket. During the few weeks of the spring selling season, plugs allow bedding plant growers to finish flats faster.

● **Mechanization and standardization:** The plug unit—even though some are square or round and come in varying sizes—creates a standard way to handle plant propagation.

Already, the tedious chore of separating bareroot bedding plant seedlings and gently placing them upright in a hand-filled flat is increasingly changing to conveyer belt transplanting lines. These production lines are fed by a flat filler, automatically dibbled and pass in front of transplanters who fill two cells at a time with plugs.

Once plugs are planted, trays pass onto an automatic watering conveyer where they're watered in before going onto a roller track for picture tags, gathering and loading onto carts for transport to the greenhouse. In the most modern ranges, trays are removed from the final conveyer by machine and placed into palletized growing benches and rolled over track in greenhouse aisles to the growing area.

Now, envision the same flat filler, but instead of feeding a human transplanting line, it feeds dibbled flats to a robotic or mechanical transplanter. And before plugs are transplanted, they're graded first for size and uniformity using optical scanning. This type of technology exists and is being adapted to horticulture.

Everything that's in this month's special issue on plugs is here and now. What's emerging just under the surface of today's technology? During the Professional Plant Growers Association meeting in San Antonio, Texas, Iowa State's Dave Koranski shared some of his views on the future.

What's coming in trays and media

Dave predicts the rise of new designs in plug trays. New trays will be breathable—they'll allow oxygen to penetrate the root system. "One of the biggest difficulties in producing plugs is oxygen," Dave says. The next big step in perfecting plug growing technology is totally artificial media. The new media will incorporate basic, needed chemical and physical properties and will be able to be reproduced. Chemical control in an artificial media must be precise. "In today's normal media," Dave explains, "you make one change in the peat and you'll get totally different results."

Computer control is here

"The next cutting edge in plugs, I believe, is totally computerized growing. It's here," Dave says. "Applying the right amount of moisture at the right time, injecting carbon dioxide (maybe even oxygen), providing the right amount and type of light at the right stage of development, computer controlled fertilizer injectors that provide chemical elements at the right time—all controlled by a computer.

"Nutritional needs change at different stages of the crop," explains Dave. "The cutting edge is not one nutritional program, but supplying what the plant needs at varying stages of growth."

Subirrigation is here

Subirrigating plugs is happening now on a limited basis, and Dave predicts subirrigation will become more widespread, but not just because it saves labor.

"Subirrigation works much better than other watering methods: all pores in the media get water and nutrients."

The future of seed

Dave predicts growers will be seeing more emphasis placed on seed varieties that perform through plug production in finishing and in the garden. Already seed breeding companies are screening new introductions in plug production prior to release. Bedding plant crop series are available that germinate quickly and over a shorter window of time.

Refined seed—physically cleaned and/or separated for uniformity or other criteria—is available in a range of crops from almost any seed supplier.

Primed seed, Genesis, the most recently introduced seed product, is revolutionizing pansy production. In primed seed, the germination process is started and then stopped. Primed seed germinates faster, more uniformly and in higher percentages over a wider range of environmental conditions.

Pregerminated seed—where the seed is germinated before the grower gets it—is the next step beyond primed seed. Somatic embryos—artificial seed— have the potential of solidifying the plug system as the method of propagation, even for crops propagated by cuttings. Simply, somatic embryos are vegetative plant clones developed through tissue culture and coated to make a seedlike product.

What does all this mean?

"More growers will have to begin to make decisions based on plugs—and they will have to be making business decisions," Dave says. Many small and medium sized plug growers will question the investment in growing their own plugs.

"

Debbie Hamrick is editor of GrowerTalks *magazine.*

Viewpoint

May 1987

by Vic Ball

Our best estimate: Bedding growers will use about 2.3 billion plugs in spring '87. Half of the 4.5 billion bedding plants grown in the U.S. will be plug started. And it's going up steadily.

Why?

We listened to some growers—that's where you get real answers. We thought their conclusions might be helpful.

S.S. Voorhees & Son, Inc., Union, New Jersey

We talked with Dale Brenneman. They operate three acres, heavily spring plants, some wholesale, some retail. Only four to five years in spring plants! They like them.

Key item: 4-inch annuals. Fifteen thousand begonias go to florists for Easter and Mother's Day at $1, plus strong demand from landscape contractors, again $1. Plus lots of other 4-inch demand (tomatoes and other flowers). At 70 cents to 89 cents wholesale. All this is grown from plugs.

Why?

"First," says Dale, "they save lots of time. Four-inch begonias, sow-to-sell with bareroot seedlings, 17 weeks. Plug begonias, plant-to-sell, eight weeks. About half as long on the bench. We buy most those plugs, and we feel we do OK cost-wise when we get 80 cents to $1 for these 4-inch plants.

"We also do lots of flats, some wholesale, some retail. They will probably be $6 this spring wholesale, 99 cents a pack retail ($12 a flat). We are all 12 packs per flat, four plants per pack. Impatiens and begonias are all from specialist plugs. Things like marigolds, we use our own seedlings, bareroot. It works fine. The plugs, again, are a lot faster, quality is better with plugs, more even. No skips to fill in. Also, a lot less seed flats to sow and worry about.

Outlook for the spring ahead? "We read somewhere that bedding plants are leveling off." (Editor's note: It doesn't say this in *GrowerTalks*!) "In fact, we couldn't supply the demand last spring. And orders are coming in fine."

Jim Gordy, J & J Greenhouses, Claxton, Georgia

Jim and Judy Gordy do 60,000 square feet at Claxton, Georgia, mainly spring plants. All wholesale. Family!

GrowerTalks on Plugs

"Flat annuals are big here. Impatiens are 36 per flat. Petunias are 48, nothing over 48. Price, $5.85, probably $6 in spring. Fall pansies, $6. All flats are from plugs except marigolds, verbenas, zinnias, and dahlias (from our own seedlings). We use specialist begonia plugs, grow most of the rest of the plugs ourselves.

"One reason we like plugs: Our transplanters do nine flats per hour on seedling marigolds, 25 to 30 per hour on plugs. One third as much labor! Besides that, it's important just to be able to keep up during the spring rush. Also, it's good not to be sowing seed all spring.

"The other big reason we like plugs: Less time! Begonias, sow-to-sell from seedlings, 15 weeks. Plugs, plant-to-sell using heavy plugs, four weeks. And space is at a premium in spring.

"We also do 200,000 4-inch for garden centers and lots for landscape contractors. Trend is up here. Price: 75 cents wholesale. They are grown mainly from plugs, some bought, some we grow.

"Our big 4-inch pansy crop in fall is all bought in plugs—it's too hot here in summer to germinate pansies. Price: 50 cents for 3½-inch.

"Then hanging baskets! We do 40,000 10-inch, mostly from heavy plugs we buy in. Lots of impatiens here. We pot five heavy impatiens plugs per 10-inch basket, plant-to-sell, as little as four weeks. And they do make a gorgeous basket!"

How's business? "We're some ahead of '86 in spite of weeks of rain. Early bookings on pot roses are up.

"We do like plugs!"

Ken Ruch, George Didden Greenhouse, Hatfield, Pennsylvania

This long-established range is very much a part of the plug revolution, too! We talked with Ken.

It's heavily flat annuals, but in a very different way.

● **Packs.** It's a brown fiber pack, 12 plants each pack, six packs per 21-inch flat. Nice, natural looking, different.

● **Wide, wide variety.** More than 250 varieties of flowers/vegetables! Seventy-five varieties of petunias alone. Says Ken, "We're unique here—and it helps command a premium price."

● **Top quality—always.**

● **Price.** "We get $7.95 wholesale for most annuals, six packs of 12 plants. We do cut begonias to nine plants per pack. Impatiens, 12 packs per flat, six plants per pack, we get $12.90 wholesale per flat.'

● **"We sold out again last spring.** Can't seem to grow enough, especially of impatiens. We will expand impatiens 40 percent this spring, half Elfins, half Accents. No real difference we can see."

Now about plugs. Says Ken, "We'd like to grow all our annuals in plugs. Much faster crop time. Example: Our 3½-inch pot impatiens are salable in as little as three weeks after transplanting from a 72-cell tray. Flats are often ready to go in only two weeks—in May. Also, we get away from loss after transplanting on peppers, vincas, and especially begonias. With seedlings, you've got to replace some skips—rarely so with plugs. And they make a more even, better quality flat. And certainly they are much faster to transplant."

● **Problems and limitations with plugs.** "We just don't have enough space to grow them all. We grow 3 million to 4 million annual plants per year, about 500,000 of these are in plugs. Some bought, mainly impatiens, a few we grow our own.

"It's hard to buy plugs of that wide list of different varieties. No one wants to grow such small quantities for us. We're aiming to grow these small lots ourselves. Probably in the long haul, we'll grow some, buy some. We are building 10,000 square feet this year.

"We like plugs. We're moving steadily that way."
Demand for bedding? "Strong. I just can't grow enough."

Carole Barton, Barton's Greenhouse and Nursery, Alabaster, Alabama

Carole does about 40,000 square feet just south of Birmingham. She is big on 4-inch annuals, some perennials, geraniums and some poinsettias.

Carole is a classic "start-up," only five years in the business. Mostly she has worked hard to develop demand for 4-inch annuals—and she seems to have won. Main customers: Landscape contractors (malls, industrial parks, etc.). "We get 58 cents for 65 cents in volume. We space, aim for quality. Garden centers are also an important customer (retail $1 to $1.59). Pansies are big in the fall for both outlets. These 4-inch annuals have become our niche.

"We got help starting up from a friendly banker and from family. I think we've hit the turnaround point this year."

Plugs are big here, too! Says Carol, "The majority of our 4-inch annuals are from specialist plugs. Really, all except marigolds—we use our own seedlings for them. All impatiens and begonias, snaps, vinca are from plugs."

Why plugs?

• **"With plugs**, we can open up, start heating our greenhouses February 1 instead of January 1. That saves a month of mid-winter fuel costs and that's big. Also, plugs produce a salable 4-inch plant easily in four to five weeks less time than seedlings."

• **"Lot less transplanting labor with plugs. Very little replanting.**
"I like plugs!"

Attention hort graduates: Carole finished in hort at Ohio State University, looked over the modest starting pay offers, decided to go it on her own.

"It's hard work, but I enjoy it." **"**

Vic Ball is editor in chief of GrowerTalks *magazine.*

Growing plugs from A to Z

January 1987

by David S. Koranski

Single-cell plant production, or plug production, has mechanized the floriculture industry. Plug production provides several advantages not found in traditional bedding plant methodology. Mechanization has made the processes of seeding, growing and transplanting more efficient; time, space and labor are optimized. Single-cell plants are more vigorous, resulting in a reduction in time needed to produce a crop, thus allowing two or more crops in one season.

There are some special considerations a grower needs to think about before producing plugs. A substantial initial investment is necessary for germination facilities, mechanical seeders and trays. The germination and growing-on of the seedlings requires a controlled environment with accurate monitoring devices to provide the proper levels of light, temperature, moisture, gases and nutrients.

Much progress has been made in plug production over the past few years. In this article, I will discuss the basic skills and cultural requirements needed for plug production, as well as present information from our latest research observations.

Seeders

Many changes in equipment currently on the drawing board will have an impact on the seeders manufactured. Mechanical seeders manufactured continue to expand in number and improve in quality each season as the machines increase in speed, seed placement and accuracy. Individual seeders already can offer the grower a variety of options. One new option will allow the operator to use smaller amounts of seed in the machine.

The seeder choice depends on the size of the operation and the types, sizes and shapes of seed most frequently sown. A grower who is just starting to produce plugs might purchase an inexpensive machine until becoming familiar with the difficulties associated with plug culture. A seeder should have the capability of handling many different sizes of plug trays. In addition, the seeder should have a calibration system that allows manual or automatic adjustments for manifold height, vacuum, pressure, vibration, tilt, rpm's and so on. Any or all of these adjustment features are a plus for that system.

It is important for growers to assess their needs for accuracy, speed, type of seed to be sown, and future expansion before making a commitment to purchase a particular type of seeder. Purchasing a seeder is a major commitment, and one needs to consider the overall system that will be used. With so many new options in the planning stage, it might be a good idea to wait one year before purchasing a new seeder.

Plug trays

The grower can choose from many different sizes and shapes of plug trays. We have found one of the most important aspects of a plug tray is its depth in relation to the air porosity of the medium in the tray cells.

A 6-inch pot containing peat and vermiculite will have an air porosity of approximately 20 percent and, thus, excellent drainage. The same medium in a 406-cell plug tray would have an air porosity of approximately 1 percent to 2 percent. The difference is the depth of the container. At least a 2-inch column of soil is needed to drain water. The deeper the cell, the more oxygen. Many growers are trying to germinate with 0 percent air porosity in small plug cells.

Germination techniques

Seed germination is one of the major obstacles challenging the plug grower. The success or failure of these germinating methods usually depends on the ability to achieve uniform control of environmental conditions. If optimum levels of moisture, temperature and light are not achieved, difficulties in obtaining a high germination percentage can be encountered. many germination methods have been used successfully. Some of the methods include controlled environment rooms, sweat chambers and greenhouses or other structures with intermittent mist.

Culture for germination

Successful seed germination is dependent on light, soil temperature and moisture levels. Light is not necessary for germination, but emergence of the radical and hypocotyl is essential for subsequent seedling growth. We have been able to directly relate the physiological disorder, water stem, to inadequate light levels. Water stem is observed as translucency in stems, mainly on petunia and impatiens. The condition develops within one to two days after germination occurs. Light levels below 400 footcandles, are believed to be a direct cause of water stem. other influencing factors include suboptimum night temperatures, low levels of nitrogen and calcium and cultivar sensitivity. Water stem, though usually fatal, can be prevented by the provision of light intensity of 400 footcandles or greater.

Soil temperature is very important for germination. uniform temperature must be maintained throughout the germination facilities. Different cultivars need different temperatures for optimum germination. A grower can use the information on seed packages as guidelines, but a high germination percentage requires adjusting these guidelines to the grower's facilities and testing small quantities of seed for germination in plug trays. Through our research, we have determined that geraniums germinate best at 70 F to 72 F; begonias, 82 F to 85 F; and impatiens, 75 F to 78 F. Optimum germination for petunias is 75 F to 78 F; salvia, 75 F and vinca, 75 F to 78 F.

Air circulation is essential for uniform temperature in a facility where the lamps are the source of heat. A recording thermometer, or thermograph, is very helpful in monitoring air temperature; soil temperature is approximately 2 F lower than the air temperature. Bottom

Table 1: Germination percentage.		
Wand hand-water	vs.	Mist
69%		88%

heat, such as root-zone heat, can provide the proper temperatures for bench germination.

The success or failure of plug germination is directly related to the moisture applied to the seed (Table 1). Too much moisture doesn't allow enough oxygen to reach the seed, which can result in desiccation and death of the seed. In most of our research studies, we have provided moisture by using fine mist with droplet size of 20 to 60 microns. This system has provided ample moisture and oxygen for optimum seed germination. The system consists of combining air and water to provide small water droplet sizes. The size of the water particles can be changed by using a regulator. preliminary studies have been conducted by using fog—5 microns. The results are similar to the fine mist. Clogging of the nozzles by hard water occurs and can be corrected by using a filter system. propagation mist has a droplet size of 300 to 500 microns, which is too heavy to allow oxygen to reach seed.

The temperature and moisture requirements are determined by the crop and can generally be divided into three regimes for three major bedding plant crops: petunia, begonia and impatiens. petunias germinate best at 75 F and with 70 percent to 80 percent humidity. Once the seed have germinated, the growing medium can be allowed to partly dry out between waterings. Begonias germinate and grow well in an 80 F and 90 F humidity regime. Impatiens germinate best at 75 to 78 F and an intermediate

Table 2: Plug growing media for petunias and begonias.
Petunias
35% Hypnum
35% Sphagnum
25% Fine-grade perlite
5% Calcine clay
Begonias
35% Hypnum
35% Sphagnum
30% Medium-grade vermiculite

moisture level. A grower installing a new germination facility should consider "zoning" the germination facility to provide different temperatures and humidities. Zoning should provide the grower with better germination percentages than would germinating in only one environment. Humidity may be monitored with devices such as a dew point hygrometer or an aspirated psycrometer.

Germination medium

One of the problems most frequently encountered in growing plugs is the provision of a suitable germination medium. Because of the small size of the container, the medium may present serious problems. These problems include fluctuations in: moisture content, aeration, pH, soluble salt levels and nutrient levels. Therefore, a desirable medium should have a high buffer capacity, high cation exchange, a high water-holding capacity, and a broad particle size distribution to ensure proper drainage. Difficulties are encountered in obtaining optimum growth for begonias, impatiens , and petunias if they are grown

GrowerTalks on Plugs

in the same medium under similar environmental conditions. A medium with 20 percent to 25 percent air porosity is excellent for petunia germination, whereas begonias require only 10 percent to 15 percent air porosity.

Perlite can be added to the recommended petunia medium to increase the air porosity. Calcined clay can be incorporated into the medium at 2 percent to 5 percent to improve buffer capacity and aeration. Vermiculite can be added to the begonia medium to decrease the air porosity, and increase the cation exchange capacity.

Quantities of nutrients that should be mixed with the medium will vary with different sources of media, the environment, water quality and crops to be grown. Commercial mixes are becoming very popular with growers because of their uniformity and convenience. Growers must realize that commercial mixes usually contain a nutrient charge. More importantly, the nutrient charge will vary from one commercial mix to another. It is extremely important to test the mix to determine its nutrient content. Dry growers can burn seedlings if the media has a high nutrient charge. In our experiments, the best pH range at the start of the germination period for all annuals tested was from 5.2 to 6.0.

Little difference in germination has been observed in different media. The major difference observed was related to airspace. Optimum air porosity is a result of good particle size distribution. A lack of variation in the particle size of the medium's constituents may contribute to poor germination and poor-quality seedlings. A medium must have the appropriate physical properties to allow it to dry partly between waterings. We have observed numerous flats in which the germination percentage was very low. Close examination of the medium indicated a lack of water drainage as the probable cause. The particles of the medium were similar in size, and they compacted with watering, preventing sufficient drainage. The seed may not have survived because of a lack of oxygen caused by a water-saturated medium. Careful preparation of the medium and testing it for drainage in different-sized germination containers should prevent serious problems. Calcined clay may be added at 3 percent to 5 percent to improve aeration.

Water quality

As it is important to know the quality of the medium, so it is important to know the quality of the water. A germination medium of top quality may be of no value if the water applied is of poor quality. Bicarbonate fluoride, chloride, sodium and boron levels will cause major difficulties if they exist in excess. Bicarbonate levels can be neutralized with sulfuric, nitric or phosphoric acid. If the other elements are present, a water purification system should be seriously considered.

Germination facilities

Germination rooms are environmentally controlled rooms or chambers in which seeded plug trays can be placed to germinate. They are designed so the trays may be stacked vertically on movable carts and rolled out of the room for observation, or for the transport of the germinated seedlings to a growing-on area. Most of the heat for the growth room is supplied by cool white fluorescent lamps. Air conditioning is usually added to the chambers to provide uniform temperatures. Seedlings should not be hand-watered. Better results are obtained with the addition of a fog or mist system, 5 and 20 microns respectively. These systems increase the humidity at plant level, which is important for optimum germination. hand watering can result in seed being washed out of plug cells. Generally, only spot watering is needed when fog and mist systems are used. Seed trays, depending on the crop, usually remain in germination rooms approximately two weeks.

The walls of a germination room can be constructed of exterior plywood. The inner compartments should be a least 3 feet wide so that carts containing plug trays can be wheeled into each section. A slanted roof is used to prevent water droplets from falling on the seed flats. When high-pressure mist or fog is not used, a false ceiling with perforated holes can be substituted for the slanted ceiling. Cool air would then be forced through the holes to help maintain uniform temperature within the room. cool white fluorescent tubes are installed between the compartments and are placed 8-inch to 10-inch horizontally or vertically from the tray. The light intensity measured at plant level should be 400 or more footcandles for most crops.

Sweat chamber

Though very similar to germination rooms, sweat chambers are designed for germination only. Therefore, seed trays remain in the sweat chamber for two to four days, or until the radical, or new root emerges. A 90 percent relative humidity is maintained by mixing air and water at 15 pounds of pressure. An 80 F temperature is provided by hot water pipes located at the perimeter of the chamber's walls. Sweat chambers provide a very controlled environment. Different temperature/humidity requirements can be obtained for different crops. Growers who have optimum control of the environment in their growing-on facilities usually have success with this method of germination. Problems arise when these young seedlings are removed from a dark, moist environment and brought into a dry, high light intensity greenhouse. These seedlings are at a sensitive stage and require gradual adjustment to the greenhouse environment. To help ease the adjustment of seedlings to the greenhouse environment, high output fluorescent tubes, or High Intensity Discharge (HID) lamps should be installed in the chamber to provide 400 footcandles of light. An additional low light intensity area of the greenhouse can also provide a controlled moisture and temperature regime before moving the seedlings into the growing-on area.

To construct a walk-in sweat chamber, exterior plywood walls can be built and lined with polyethylene. A high-pressure mist or fog system should be placed in the middle and at the top of the chamber. The moisture droplet size should be five to 80 microns to allow adequate oxygen to reach the seed. The best uniformity of moisture distribution can be attained by very little air movement. Too much air movement can cause uneven distribution and drying in some areas. Uniform moisture distribution is one advantage of the sweat chamber. heat is added by passing hot water through $1^1/4$ inch pipe placed in the chamber around the inside walls. This method allows a uniform 80 F (±1 F) to be maintained.

Germination on benches

Germination on benches allows the advantage of close inspection of the plug trays. Root-zone, perimeter or overhead heating systems should emit approximately 70 Btus per square foot to provide optimum temperatures. Overhead fog or mist systems supply the necessary moisture. It is difficult to use irrigation booms for smaller seed. The large water droplets inhibit oxygen penetration to the seed. Some growers use capillary mats on the bench, especially if using root-zone heating without fog. The mats act to supply moisture to the plug trays and reduce the drying conditions of root-zone heating.

HID lamps provide supplemental light energy for seedling development after germination. Maintaining uniform conditions at plant level is a major task, but necessary for optimum germination and subsequent growth and development of the seedlings. Depending on the crop, plug seedlings may remain on the germination benches for one to four weeks before being moved to a growing-on area. Containerized and movable benches are becoming very popular with growers and work well for germination. Containerized benches can be easily moved to different growing regions on special tracks in the greenhouse without the need for handling the individual trays.

The germination facility a grower should use depends on the size of the operation and the feasibility of achieving uniform environmental control in the germination area. The key is uniform temperature and moisture control. Small growers, unless specializing in plugs, often do not have adequate greenhouse bench space to germinate plugs. The solution to this problem is to take advantage of stacking the plug trays vertically in a special germination room or sweat chamber that has controlled humidity and temperature. Growers with older greenhouses should consider germination rooms or sweat chambers. Older facilities usually are not designed for the precise control needed for temperature and humidity. A large grower may choose to germinate plugs on greenhouse benches provided that environmental control is possible. Benches allow easy access to the seedlings for care and observation.

Temperature, moisture and light for growing-on

When seedlings are removed from the germination area to a growing-on area, they must be allowed time to adjust to the different environment. The first move, to Stage A, is to a greenhouse that can provide 70 F to 80 F soil temperatures for two to five weeks. The plug seedlings should be watered daily, preferably with overhead mist at 60 F to 70 F. Uniformity of the moisture application will determine the uniformity of the growth and development of the seedlings. Turning or rotating flats also may be beneficial.

Field observations have shown that petunia seedlings placed immediately under supplementary HID lamps grow and develop much more rapidly, and they flower approximately one to two weeks sooner. The light intensity should be from 450 to 1,000 footcandles under HID lights. Four crops that we have tested have been shown to benefit from supplemental lighting.

Optimum growth has been observed when impatiens were provided with two to two and one-half weeks of supplemental lighting; petunias, two weeks; begonias, two and one-half to four weeks; and geraniums, four weeks. Growers should keep careful observations on the crop. Chlorosis or bleaching is an indication that too much light is being supplied.

Light intensity and quality are still important factors for seedling growth in Stage B. A higher light intensity, 4,000 to 6,000 footcandles, encourages compacted growth. Impatiens are sensitive to high temperatures and high light intensity. They benefit from being shaded beginning in mid-April. This is especially important once they begin to flower.

In addition, the growth of seedlings can be controlled by the amount of water they receive, as well as by their ambient and soil temperatures. For slower growth, the amount of moisture applied can be reduced. However, depending on the plug size, some seedlings cannot be allowed to lose turgidity. Stunting and flower delay may result from the lowered temperatures. For rapid, lush seedling growth, moisture availability should be constant and combined with 75 F temperatures.

Growth regulators

Seedlings growing under conditions of low light intensity, temperatures above 75 F, and high moisture are likely to elongate. Application of chemical growth regulators is necessary when the environment can no longer be adequately controlled, as in late spring.

The application time for the growth regulator is one of the key factors in the success of growing quality plug seedlings. The growth retardant should be applied before the seedlings begin to elongate. Plug seedlings can be treated at earlier stages of growth and development than non-plug seedlings. To achieve the best results with chemical growth regulators, the first application should be made during Stage A, when the seedlings first begin to elongate, and are 1 to 2 centimeters (1/2 inch) in diameter, or at the first or second true-leaf stage. Because some compounds may inhibit root development, the root system should be examined to ensure that there is active growth before growth regulators are applied. Seedlings also should be of uniform size at the time of treatment to obtain consistent results.

The method of application to plug seedlings is similar to that for cell packs. Fine droplets are desirable for uniform coverage and penetration of the foliage. The amount of material applied (milligrams of active ingredient per flat) should be monitored and controlled so that the same amount is being applied to each flat. We believe the most suitable time to apply chemicals is approximately two hours after sunrise on a cloudy day because the stomates in the leaves are open to a maximum and can allow adequate penetration of the chemical.

Nutrition for growing-on

Nutrition is of special importance in the growing-on stage. We have conducted some preliminary experiments that have shown excellent results with feeding petunias two days after sown with 50 parts per million of a 20-10-20 fertilizer. The experiment was conducted by using 20, 50 and 75 parts per million at every irrigation. The 50 parts per million feeding provided the best growth and development.

Soil tests should be taken at least every two weeks. The most common problems encountered from soil tests have been high soluble salts, high nitrates and low phosphorus. Several thorough irrigations with clear water will leach the salts and nitrates. Phosphorus can be incorporated into the medium as superphosphate (0-20-0). But, phosphorus has also been detected at excessive levels. This element can tie up iron, which can cause iron deficiency. Excessive phosphorus can be leached out of soilless media with several thorough irrigations.

Low pH can tie up magnesium and calcium, causing a deficiency of these elements.. Magnesium deficiency usually is evident as yellow chlorosis on leaves in the lower portion and middle of the plant. Marigolds and petunias seem to be more susceptible to magnesium deficiency than most other annuals. Raising the pH and adding magnesium sulfate (Epson salts) usually corrects this problem. High pH (6.8 or higher) is a common result of continuous use of fertilizer solutions containing a high concentration of calcium nitrate. Iron deficiency may result from this higher pH, appearing as the classical interveinal chlorosis. Even though iron may not be lacking in the soil, high pH makes it unavailable to the plant. Iron chelate and iron sulfate can be used, with caution, to remedy the problem. A fertilizer mixture with minor elements and a relatively high percentage of nitrate nitrogen is used to maintain optimum nutrition and optimum pH.

Timing and scheduling

One of the many advantages of growing plugs is the reduction in the time needed to produce a crop. Timing for plug production must piece all stages of growth together with the environmental conditions for a smooth-flowing schedule. Most seedlings need to remain in the chamber seven to 14 days. If left in the chamber too long, the 75 F to 80 F chamber temperatures and relatively low, 400 to 500 footcandles light intensity will cause stretching and non-uniform development. field observations have shown that the earlier seedlings are removed from the chamber (from the first sign of germination to when the cotyledons begin to unfold) and placed in Stage A (a 75 F greenhouse equipped with supplemental HID lighting), the more rapidly and fully the seedlings develop. In addition, stretching is greatly reduced. if seedlings must remain in the chamber for any extended time, the temperature should be lowered as much as possible and the light intensity increased to help prevent stretching.

The warm, 70 F to 75 F, temperatures of Stage A serve to acclimatize the tender, newly germinated seedlings and to provide an environment that, when supplied with moisture and fertilizer, encourages rapid growth and development. Most crops remain in Stage A for two to three weeks or until the first set of true leaves begins to appear. Begonias may remain much longer, even up to the time of transplanting.

As the true leaves begin to develop, seedlings may be moved to Stage B, a cooler (55 F to 60 F) area to "hold" until needed, or a warm (75 F to 80 F) area for continued rapid, succulent growth. We have found that the flowering of petunias may be delayed by subjecting plants to temperatures below 60 F before the plants have initiated flowers. Seedlings remain in Stage B until transplanted, approximately two to three, or more weeks.

Once transplanted, the seedlings can again be "held " at 55 F to 60 F or encouraged to develop and flower in approximately two weeks by providing 70 F to 75 F temperatures, moisture and fertilizer.

In summary, growers seeking to enter plug production or upgrade their current system should investigate all equipment and structures options. Success in plug production remains dependent on attention to detail. This article presents only a few of the requirements that we have observed. Growers must apply this information to their own greenhouse growing environments.

,,

David S. Koranski is professor of floriculture at Iowa State University, Ames.

The exploding world of plugs!

January 1987

by Vic Ball

First off, the five points developed here are based on the two-day Ames Plug Conference, and just as importantly, on a thorough briefing by Dave Koranski—on work being done in his lab at Ames (supervised by Tricia Nicholisen).

Here follows then, as briefly as possible, these important new approaches to plug production. Things of practical help to the grower.

For the record, a major gathering of plug growers, the National Plug Production Conference, was held at Iowa State University, Ames, Iowa, November 5 and 6. About 380 growers paid $200 each for these two days of seminars (including banquet and lunch). Sessions covered a variety of plug topics. Sessions were certainly interesting and well participated in by the growers present. The whole affair was a great credit to Dr. Dave Koranski and his able staff. There will be a summary of talks published within a few months—available from Dr. Koranski, Department of Horticulture, Iowa State University, Ames, Iowa, 50010.

1. Fog: A major help

Important: Plug flats started off with fog grow notably faster and better than watering by hand or even large particle nozzle irrigation. There is a place for large particle irrigation as plugs mature, but fog is clearly superior in the early stages. Main reasons: Fog supplies constant, even humidity, which is critical to soaking seed coats and starting germination. Second, large droplet irrigation tends to overwater and overleach the tiny plug soil ball. Petunias and impatiens are not swamp plants! Roots need oxygen.

• **Fog—small droplets**: Five micron particle size (.0005 millimeters) is pure fog—like being in a cloud! Fog provides excellent humidity and a bare minimum amount of water on the crop.

• **Large particle irrigation**: Three hundred to 500 micron droplets are "large particles." It's the old traditional propagating nozzle.

An easy way to differentiate: Hold your hand, palm open and toward the nozzle. If lots of water is instantly running off your palm, that's large particle! Five hundred microns. If your palm slowly becomes a bit moist, that's fog. Five microns.

• **Fine mist**: Dave suggests an in between particle size, from 20 to 60 microns. The fine mist particles do essentially the same job of creating humidity, without water, as you get from fog, and at substantially lower cost. Less air pressure for one thing.

How to generate fog (or mist)

First, there are two basic ways:

• **Hydraulic atomizer**: It starts with a small tank—containing, in our case, water. The water, or whatever fluid is put under pressure, is piped out through a nozzle and it makes either fog or "coarse particle" mist. The common Hudson sprayer is a good example. The pressure on the fluid can be anywhere from 30 pounds to 50 pounds per square inch (psi) or up to 1,000 psi. This system can produce either coarse particle mist (500 microns) using low pressure and an appropriate nozzle, or it can produce fog, now using perhaps 500 or 1,000 psi and a fog-type nozzle.

• **Air atomizer system**: The fluid (water or whatever) is piped into the nozzle area, compressed air is also piped in, mixing occurs and the nozzle produces small droplets. It's a different system from the hydraulic atomizer, not necessarily better or worse, each fits certain applications best.

Where and how to get the equipment

Again, there are several ways to go. Among them are, in alphabetical order:

• **AGRITECH**—P. O. Box 577, Broadway, North Carolina, 27505, (919) 258-9113. These folks make the 3 foot round cylinder device that rotates, and again produces fog in greenhouses.

• **Atomizing Systems**—Dalebrook Industrial Park, 1 Hollywood Avenue, Ho-Ho-Kus, New Jersey, 07423, (201) 447-1222.

• **Baumac International**—(Micro Systems), 1500 Crafton Avenue, Mentone, California, 92359, (916) 332-1318.

• **Mee Industries**—1629 South Del Mar Avenue, San Gabriel, California, 91776, (818) 288-4650. Specialists in fog generating for various applications.

• **MicroCool, Environmental Cooling Concepts, Inc.**—653 Commercial Road, Palm Springs, California 92262, (619) 322-1111.

• **Spraying Systems, Inc.**—North Avenue and Schmale Road, Wheaton, Illinois, 60188, (708) 665-5000. Ask for Jim Pelej. Spraying Systems are nozzle specialists (2,500 models), they offer self-contained foggers/engineering on larger jobs.

About cost: for a rough idea, a 30 foot by 100 foot greenhouse, to be fogged with let's say 40 to 60 micron particles, using systems from such firms as Mee or AGRITECH, will cost on the order of 75 cents to $1 per square foot of ground covered.

2. Glassy stem—a new problem

Probably not really new, just newly identified.

Symptoms: The "stem" of a newly sprouted seedling (¼ inch or so) becomes glassy, translucent. Soon (several days), the seedling dies. It affects mainly petunias and impatiens a bit (so far), typically 10 percent to 20 percent of a tray. It is apparently a physiological problem, not a disease.

Control: Dave's work clearly demonstrates that applying 450 to 500 more footcandles of supplemental light 24 hours a day, starting from the time of sowing, greatly minimizes or eliminates the problem. Fluorescent or HID light will do the trick. Dave's only other comment: Varieties differ importantly in their susceptibility. By the way, begonias are not affected by this problem.

As a practical matter, glassy stem can, when it appears, cost you 5 percent, 10 percent or even 20 percent of your petunia seedlings. Five hundred footcandles of light for 10 days almost eliminates the problem. And besides, the light provides some other important benefits. So, why not light?

For the record, we tried for 15 minutes to pick out the "glassy stems" in a tray of seedlings. It's almost impossible with the naked eye. Under a hand lens or equivalent, and with a few minutes practice, it's not hard to spot them. Glassy stem was the subject of a lot of discussion at the Ames meeting.

Dave showed a slide with numbers on the effect of light on glassy stem during his presentation:

Light level	Glassy seedlings (%)
400 footcandles	0
200 footcandles	22
100 footcandles	59

3. Refined seed—pro and con

The subject got a good airing at the conference. Four major producers of plugs reported their experiences with refined vs. normal seed. Quite candidly. The session was preceded by a report by Carol Lewnau of the Iowa State University staff. She had plainly done a comprehensive job of testing refined seed vs. normal, well replicated, etc. To capsule her conclusions, some varieties showed perhaps 4 percent to 5 percent higher germination, some did not. Some varieties germinated 5 percent, 6 percent or 8 percent faster and others didn't. Some varieties showed less of what she called runts. Over and over, her comment was that "varieties differ widely." Refined Happiness petunia germinated 20 percent better than regular. Others showed no difference. "Refined seed generally were larger," she says. Dave made the point that this study was done with the refined seed available two years ago—and that undoubtedly there has been improvement since then.

The important conclusion from all this would seem to be that the refined seed was always as good as, and often better than, regular.

The growers on the panel in general seemed to have reached about the same conclusions. Allen Smith, Frank Smith and Sons, Michigan: "I've seen dramatic improvement on some varieties, very little on others. You've still got to do the germination job right, and if you do, you will get perhaps 5 percent to 10 percent better germ on some varieties."

John Williams, Tagawa Greenhouses, Denver, again major plug producers, reported, "Seed on some varieties emerges more evenly. We use refined seed wherever it is available. In fact, we don't grow many varieties where it is not available. Mainly it seems to provide more uniform trays. Also, seed seems to be cleaner, works better in our seeders.

"We need better verbena germination." Gary Spear, Charlotte, North Carolina, again, major plug producers, "We don't use refined seed at present—we just don't see the need. We did tests in 1985, didn't see that much difference. Refined seed was more even. The dahlia seed especially was cleaner."

Preston Hopkins, John DeWinters, Hudsonville, Michigan: "The combination of refined seed and HID light has enabled us to make great strides toward even plug flats in less time. We didn't see the misses until we had plugs—then we looked at those empty squares. I think the seed companies are making great strides. By the way, plugs are sure the way to go. We are able to ship in as little as four weeks after transplanting, gives us a lot more double crop capability."

Dave suggests that uniformity is perhaps the major improvement that he sees in refined seed. He feels that growers should be able to move flats out of the growth chamber faster because of this.

Conclusion: Heartening progress here, major seed companies still hard at work on the matter.

4. New feeding regime—major improvement

General industry practice (our observation among most growers) starts feeding plug sheets about 10 to 12 days after sowing. Tests at Iowa State University show clearly that if feeding is started on day 2 (after sowing) that the grower will turn out finished plug trays in about seven to 10 days less total time! That's a big difference in any range.

The comparisons we saw at Dave's lab during our tour also showed clearly more even stand where feeding was started sooner. Dave reported that begonias showed an even more dramatic difference than the petunias that we saw.

Trays were treated with 25 and 50 ppm of nitrogen fertilizer. The 50 ppm trays were clearly better than the 25.

The nitrate levels described above were applied with all irrigations. By the way, this feeding is generally done with boom irrigation with 300 to 500 micron particles. Including liquid fertilizer with even, fine mist (20 to 80 microns) is at best "iffy".

5. Winter light—the problem and some answers

These paragraphs are straight from Dr. Ted Tibbitts, University of Wisconsin—his presentation at the Ames plug conference. In 20 minutes he clarified so much. A practical guy!

First, why supplemental light? Answer: In our northern growing areas, December light is a shocking 21 percent of July light. See chart on solar radiation.

This sharp reduction is, of course, the result of the sun being very low in the southern horizon (available light is spread over a wide area). Also, cloudy weather so prevalent in our Midwest, worse as you go East. Also, of course, daylength itself is reduced from 15½ to about nine hours daily.

This severe reduction (to 21 percent of July) is also aggravated by:

• **An estimated reduction of 50 percent** from dirty glass or dirty poly, roof bars, ventilators, etc. So now the 21 percent of July outdoor light shrinks to perhaps 10 percent or 12 percent of July light. Small wonder that plant growth, November through December is so miserable. And the Great Lakes' region (and further East) is even a bit worse. This does, of course, improve as you go south (longer days, clearer days) and southwest, even more so.

All of this bears on any winter—early spring Northern greenhouse crops. Certainly it affects winter pot crops, pot mums, etc. Winter light shortage affects plugs! Much of the growth is done at lower light times of the year. More later on this.

Solar radiation[2]
percent of maximum month

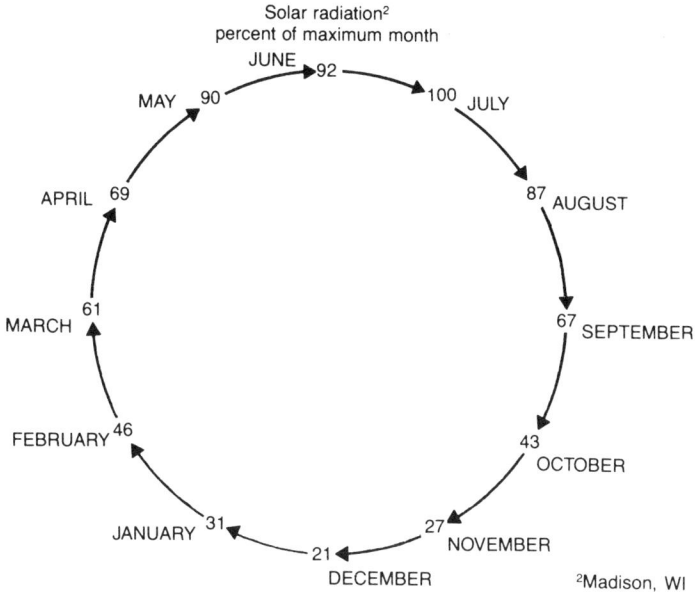

JUNE 92
MAY 90
JULY 100
APRIL 69
AUGUST 87
MARCH 61
SEPTEMBER 67
FEBRUARY 46
OCTOBER 43
JANUARY 31
NOVEMBER 27
DECEMBER 21
[2]Madison, WI

A diagram of radiation received over the year at Madison, Wisconsin. It's a Ted Tibbetts' chart that dramatizes the difference in summer versus winter light. The units are based on accumulated light energy. For example: December is 21 percent of July radiation.

Which light source?

Ted, who obviously knows this subject, said without batting an eyelash, that high pressure sodium is by far the most practical, best way for growers to light. He reports that most of the light is in the yellow band, 550 to 650 nanometers—not perfectly spread out, but clearly high pressure sodium produces "the most photosynthetically usable light of any artificial light source." Much more so than fluorescent and incandescent. High pressure sodium yields about 10 percent more photosynthetically usable light than low pressure sodium. Also, most of the wattage put into these lamps ends up as heat in the growing area. Very little is really used in photosynthesis. Which does mean a helpful reduction in fuel cost—which many growers have already learned. Two commercial high pressue sodium bulbs: Lucalox (GE) and Lumalux (Sylvania). Ted also points out that high pressure sodium bulbs have a longer life than other light sources. He estimates about 24,000 hours normal life—and "you lose only about 10 percent of the light during this 24,000 hour span. Fluorescent bulbs lose about 30 percent. With low pressure sodium, you do not lose intensity, but you must raise the wattage as the bulb ages to offset the loss in light.

"I'd use high pressure sodium."

"Low pressure sodium produces more brightness for people, but high pressure sodium is more efficient for plant growth—more growth per watt."

Again Ted: "High pressure sodium may tend, in certain crops, to cause somewhat more stretched growth, longer stem growth on seedlings—but only if there is no sunlight reaching the plants. It might be appropriate to mix high pressure sodium with metal halide." But Ted also added: "High pressure sodium does a very good job on plugs in greenhouses along with some sunlight." Ted reports that "lots of plugs do grow very well in growth chambers with only high pressure sodium."

Reflectors—also important

Again, the practical scientist: "Without reflectors, about half of the light energy from the bulb is lost—it goes out and up. So, good reflectors you must have. A negative: The larger the reflector, the more sunlight is kept from crops by the reflector itself. So obviously, design of reflectors is critical." He mentioned several companies working to make smaller, more efficient reflectors: P.L. Light Systems, Energy Technics, and Sylvania. Also Cannon, Inc., St. Danden, Ontario, makes a "good reflector."

Also, "It's important for a reflector to distribute the light uniformly over the plant area. As you go higher, light tends to spread less uniformly. In fact, the reflector should be designed to accommodate the greater height of these bulbs. Incidentally, fewer large bulbs means less shade from reflectors."

How much light?

Again, in generalities: Says Ted, "Normally from 250 to 1,000 footcandles is a practical range, but the amount of light used obviously must relate to the available sunlight. Also, of course, it must be kept in tune with what the grower can afford to install. Example: A Northern grower might use 500 footcandles in December, 250 in February."

Now, why light plugs?

A very long story, a very short punch line—several key reasons:

● **Especially during low light** December and January and in northern areas: 500 footcandles 24 hours per day will simply produce a larger, heavier, better plug in less time. And that's a compelling reason for doing it! Better quality and less time is really the name of the game. And that's the reason many specialist plug producers today are already using supplemental light.

● **Many crops flower earlier** if lighted during the plug stages: We recall tests done in spring 1982 by Herb Verbeek, Mills River Plant Growers, Ashville, North Carolina, on this point. Example, petunias. Clearly, the HID caused a week or two earlier bloom in the packs. Another good reason for lighting. Probably most other crops would respond similarly. No data.

● **Glassy stem**: This can be virtually eliminated by use of 400 to 500 footcandles 24 hours a day for the first several weeks of plug growth.

Lastly, how long to light plugs? Our best reading on this to date: Light throughout the normal growth phase of plugs—perhaps four to six weeks from sowing. Here's a "horseback" table for spacing of bulbs:

	For 250 footcandles	For 1,000 footcandles	Height above crop
400 watts	8 foot centers	4 foot centers	8 feet
1,000 watts	—	7 foot centers	12 feet

Again, 1,000 watt bulbs make less reflector shade vs. 400 watts—there are fewer of them.

About light measurement: Ted's first recommendation is the GE footcandle meter, about $50. "Fairly accurate for sunlight. Not good to read high pressure sodium, the light quality confuses the meter. Suggestion: Have it recalibrated by your extension agent for reading high pressure sodium." We have reports that this meter does not measure light accurately if the light approaches at an angle.

These meters lose sensitivity over years of use, especially if exposed to full sun. Better recalibrate them occasionally.

Several other points on plugs

● **Putting plugs on hold**: Growers frequently have occasion to just keep plugs sort of in reserve after their normal growth period. We heard several times during the Ames Conference that Bonzi is very effective. It "keeps the plugs short—and they tend to become somewhat firmer—a bit harder." To which Dave Koranski adds the comment: "The whole plug concept is designed around keeping the plant in active growth through the plug phase and immediately then into the final pack. Plugs can be held, but you pay a price in uneven flowering and hardening of root system."

New seeders

Lots of improvement in seeders, mainly minor refinements so far. The only important new thrusts are the drum seeders. The drum seeders use a rotating cylinder with holes over the entire surface. Seed is held against the surface of the cylinder by vacuum and released as the drum rolls across the plug tray. The drum seeders are very fast, very accurate, and expensive. So far mainly adaptable to the larger grower. Greiling's (Natural Beauty), biggest plug producer in the United States today, is using mainly drums.

• **Water quality for plugs**: Dr. John Peterson, Ohio State University, reported on this at the conference. Sorry we missed this but, in a word, certainly high bicarbonate and high pH water sources are not the greatest for plugs. The other important point: Water temperature. As with most crops, using cold water, which means cold soil much of the time, is not great.

• **A point on feeding plugs**: We learned several years ago that pot mums fed with a combination of slow release fertilizers (Osmocote, Magamp) and used with soluble fertilizers produces clearly better growth than either one separately. It's a classic case of synergism. Roy Larson gave an impressive report on response of plugs to mixing Osmocote in the plug mix. Dave Koranski's research documents clearly the response of plugs to feeding—from the second day after seeding. The obvious other opportunity: To mix a light application of one of the slow release fertilizers in with the plug mix—and also to apply liquid fertilizer as discussed in the above notes. Roy Larson's recommendation: "Eight pounds of Osmocote per cubic yard—incorporated in the mix before sowing (don't mix it too far ahead)."

• **Perennials**: There were several talks on the program, some encouragement of use of plugs for perennials. Problem: Somewhat slow and irregular germination of many perennials, and many which simply are not reproduced from seed. But again, plugs seem to be taking an important place in perennial production.

• **About growth chambers**: They're great for scientists who need carefully controlled conditions to compare techniques. They're also OK for the smaller grower who doesn't want to have to provide temperature, humidity, and light levels for plugs over a whole greenhouse.

Our own opinion: Any grower who does as much as a full house of plugs (25 foot or 100 foot or so) probably will do better to just convert the house over the plug conditions. That means, if at all possible, installing some sort of fog, 40 to 60 micron particles, certainly some sort of rootzone heating to permit accurate control of soil temperature, also better if you have supplemental light.

Most growers doing this also have boom irrigation—especially on later stages of growth, probably not on the initial week or 10 days. All this should be aimed at the first week or 10 days after sowing seed. The HID light would be helpful after that, but fog and higher soil temperatures are no longer needed. Once seedlings are well established, better move them under cooler temperatures, lower humidity. But they are still better off with supplemental light.

"

Vic Ball is editor in chief of GrowerTalks *magazine.*

Seed and Germination

Quality seed and Stage I—Your best start toward turning a profit with plugs

December 1989

by Paul Karlovich and David Koranski

One of the most important details in plug production is the environmental conditions in Stage I. At Iowa State University we've been studying this stage for several years. Stage I is arguably the most critical stage of plug production. What are the most important factors to consider when thinking of Stage I?

All-important seed metabolism

Water uptake. When a dry seed is subjected to a moisture source, the most obvious event that occurs is rapid water uptake. Flower seed initially absorbs water very quickly and within four to eight hours has absorbed most of the water necessary to germinate. After this initial uptake, a lag phase in water uptake occurs, lasting from hours to days. This is a "re-organization" period where events necessary for seed germination to proceed are occurring.

All seed, viable, dormant or dead, have similar water uptake patterns at this point, but only viable seed show another period of water uptake coincident with radical emergence from the seed coat.

Respiration. Another metabolic event that occurs almost immediately after water uptake begins is respiration. Within minutes of placing seed in a wet environment, the seed is metabolically active. This activity is initially disorganized, but usually by the time the seed reaches the lag phase of water uptake, quality seed are metabolically organized. The key here is that *quality* seed are organized. Low quality seed may take longer or may not organize at all, resulting in low germination uniformity or poor germination.

Table 1. Effect of soil moisture level on germination percentage of impatiens seed placed on top of or buried in the medium.

Cultivar	Moisture	Germination Top	Germination Buried
A	Driest	70%	35%
A	Intermediate	67%	11%
A	Wettest	59%	2%
B	Driest	93%	66%
B	Intermediate	94%	35%
B	Wettest	93%	15%
C	Driest	97%	72%
C	Intermediate	97%	37%
C	Wettest	98%	14%

Membrane reorganization. One of the major differences between high and low quality seed is the ability of seed membranes to organize and be functional. As water uptake begins, seed membranes are not organized and cellular components can leak out of the seed. Of importance here are carbohydrates and sugars, which are important not only for the seed during early germination, but also serve as a source of food for pathogens if they leak from the seed. Quality seed leaks less and so conserves valuable energy reserves for germination and provides less food for pathogens.

Proper environment is key

For optimum germination, proper environment is critical. The most important environmental conditions, temperature and moisture, are controllable and usually make or break efforts to achieve rapid, uniform germination.

Fresh Weight (mg)

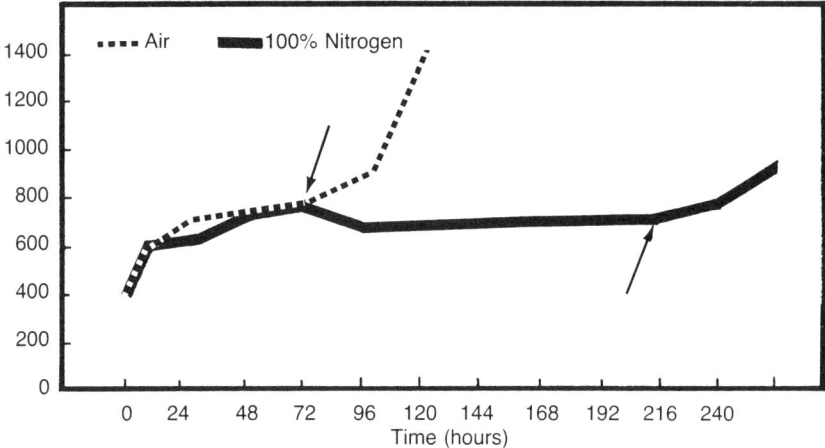

Fig. 1. Fresh weight gain of 400 impatiens seeds in air or in nitrogen gas 120 hours before transfer to air. Arrows indicate the onset of germination.

Temperature. It is essential to provide the *optimum* temperature. Most seed germinate over a range of temperatures, but for plug production, suboptimal temperatures are not acceptable: They result in slow, irregular germination or no germination at all.

Hand-in-hand with proper temperature is temperature accuracy and uniformity. Check and calibrate all thermostats regularly to be sure that "what you see is what you get." The temperature in the germination area should vary by no more than plus or minus 2 degrees F.

Germination chambers can achieve uniform temperatures, and this is one reason we recommend them. Do not underestimate the importance of uniform conditions. A few degrees variation significantly changes germination characteristics of most seed. A uniform, but non-optimum temperature can play havoc with production schedules.

For germinating most flower seed, chambers at three temperatures would be best: 78 to 80 degrees F, 70 to 72 degrees F and 64 to 66 degrees F. These temperatures cover the optimum ranges for most commonly germinated flower seed. If three temperatures are not possible, then two temperatures, 74 to 77 degrees F and 65 to 68 degrees F, are needed. With only one temperature, you will not be able to germinate a full complement of flower seed.

Moisture. Equally important to temperature are moisture conditions surrounding seed. As with temperature, uniform, reproducible and controllable moisture conditions are important in Stage I. A seed actually needs very little water, quantity-wise, to germinate. Even a moderately moist medium provides enough water to germinate.

Notice that the 400 milligrams of dry seed in Figure 1 began to germinate when their fresh weight was approximately 700 mg. Roger Styer and his group at Ball Seed Co. have germinated impatiens seed on dry media in a fog chamber and seed absorbed all the water they needed from the air! What is more important than the water quality available to the seed is the water balance in the seed.

Flower seed are small, need little water quantity-wise, and absorb water rapidly. On the other hand, flower seed, because they are small, lose water quickly (to a dry atmosphere) and do not have much water to lose. When a seed is desiccated during germination, timing and coordination of metabolic events are disrupted. The result: poor and irregular germination. Most water use in germination is to maintain the high humidity necessary to assure optimum seed/water balance.

A further complication to the moisture picture is caused by too wet conditions, particularly when seed gets buried in the medium. Excess water can limit the oxygen supply to the seed and upset germination timing. Research on impatiens at Iowa State University has shown that high quality seed buried in medium have little chance of becoming quality seedlings. But, seed germinating on the surface of the plug medium are not affected by wet conditions.

Again, the key is high quality seed; we have evidence suggesting that low quality impatiens seed is adversely affected by very wet conditions. Our advice: Buy high quality seed and handle it properly. This means being certain that seed isn't buried in the medium. Fortunately, it doesn't appear that current plug handling techniques, when properly applied, bury seed. Using a light covering of vermiculite is not analogous to burying seed in the medium.

Germination facilities. Maintaining a correct and controllable environment is the key to success in Stage I. Germination chambers provide the most stable, reproducible environment. Not all growers have the luxury of such chambers and are successfully germinating in the greenhouse.

For greenhouse germination, covering seed with a porous, non-oil-based cloth to control moisture around the seed. This is more satisfactory than covering seed with vermiculite because of the lack of uniformity of vermiculite and the difficulty in applying it uniformly to the plug tray.

Buy quality seed

Buy your seed from a reputable company. Bargain-priced seed may not end up being a bargain. Numerous specialty seed usually are higher priced, but may be worth the investment. A relatively new specialty seed is primed seed. For crops such as pansies, priming significantly increases germination speed and uniformity.

Store seed properly. Properly storing flower seed assures high quality as long as possible. Store seed in a refrigerator. Hermetically sealed foil packs can be stored as they are, but once opened, place them in a glass jar with a tight fitting lid. Mason jars are a cheap and readily available storage vessel. Do not open the jars more than necessary when in storage.

We do not recommend storing seed for more than one year. The best advice is to purchase the amount of seed necessary for one growing season. Primed seed must be used by the expiration date listed on the seed package.

Are you intelligently producing plugs or has constantly pushing for increased production and profit blinded you to the dangers? Plug production is not hocus pocus for a select few who know the magic words, but utilizes technology and procedures usable by anyone committed to attention to detail. Providing the proper environment conditions to high quality seed during Stage I is a good step toward producing quality plugs.

"

Paul Karlovich is a recent Ph.D. graduate and Dave Koranski is professor of floriculture, Iowa State University, Ames.

How to get your germination rates on the rise

December 1989

by Roger C. Styer and Shawn Laffe

Vinca, verbena, pansy and fibrous begonia—why are these crops so difficult to germinate? Seed quality, dormancy and disease can cut down germination in the plug tray. Your sure bet for success is to monitor temperature and moisture in Stage I.

Many growers have difficulty with one or more of these four crops: vinca, verbena, pansy and fibrous begonia. Plug germination may range from 10 to 90 percent and may differ greatly with varieties. To overcome these germination problems, growers commonly double-seed vinca, verbena and begonia or will plant many more plug trays than they need of these crops. If growers continue to have problems, it would be more profitable for them to buy finished plugs from another source.

Seed quality is a big problem with verbena, vinca and pansy. Due to the method of seed production, seed quality of vinca and verbena may vary greatly from year to year. Seed dormancy and disease may reduce germination, both in the laboratory and in the plug tray.

Begonias are a very small and very fragile seed. Seed can be easily crushed or damaged, so protective vials are used for shipping.

Different varieties, series and colors will have different rates and percentages of germination. Majestic Giant pansies, for example, traditionally germinate faster and higher than Crowns, Crystal Bowls or Maxims.

Environmental and cultural requirements are very important and very specific for each of these four crops. Our group at Ball Seed, along with Dave Koranski at Iowa State University, has conducted extensive research on plug germination conditions for vinca, verbena, pansy and begonia. We present specific methods for Stage I and Stage II germination and growth, along with some of our research results. We also show you how to improve plug germination with the use of specialty seed, such as Genesis Seed.

Increasing vinca germination rates

In Stage I, soil temperature must be maintained at 80 degrees F until the initial root or radicle comes out. If temperature is 75 degrees F, total germination and speed will be reduced. Soil temperature of 70 degrees F will result in very poor germination.

Moisture should be high in Stage I (95 to 100 percent) until the radicle emerges. At that point, moisture **must be** reduced (75 to 80 percent). Achieve this reduction by allowing the medium to dry **slightly** between waterings. If too much moisture is added during this stage, seed germination will be reduced considerably.

To help retain the proper amount of moisture without overwatering, we use a light covering of coarse vermiculite (No. 2 grade). This covering allows oxygen to get to the seed while still maintaining moisture around the seed. Some vinca varieties also need darkness to germinate, which the coarse vermiculite covering provides. We obtain highest and fastest vinca germination when we water plug trays after sowing and then apply the coarse vermiculite covering. This may be due to not forming a more solid covering with the vermiculite and suffocating the seed during germination.

In Stage II, lower temperature to 75 degrees F and light to 400 to 700 footcandles. Vinca does not need to be fed until cotyledons have totally formed. A low feed of 50 ppm N from 20-10-20 every other watering is sufficient until true leaves emerge. Keep medium pH between 5.5 and 5.8. Soluble salts should be less than 1.0 mmhos/cm^3 (2:1 extraction). Vinca are not heavy feeders and do not tolerate high soluble salts in the plug tray.

Many plug growers use fungicide drenches at the time of sowing. We never had to use them as long as we follow the above procedures, particularly reducing the moisture in Stage II and using coarse vermiculite as a cover.

Genesis vinca seed is available and will improve the speed, uniformity and percentage of germination. Note the reduction in total germination of both Genesis and standard vinca seed when plug trays are germinated on the greenhouse bench. This reduction is due to not having the soil temperature high enough (80 F) or uniform enough during germination as well as using a stationary mist system to maintain moisture, which tends to provide too much moisture.

Genesis Vinca Plug Germination
Chamber vs. Bench - Stage 1

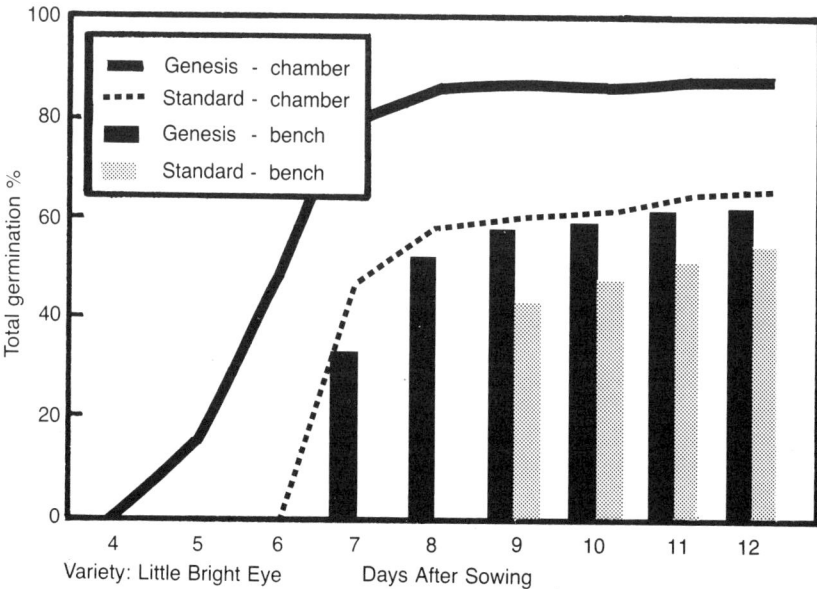

Variety: Little Bright Eye Days After Sowing

Single seeding with verbena? You bet!

This crop germinates best when soil temperatures are between 75 and 78 degrees F. However, the **most** important factor for good plug germination is moisture. Verbena requires about half the amount of moisture during Stage I as other crops. If verbena remains too wet during Stage I, poor germination will result. Increased mold growth on seed is also a problem at high moisture. We cover verbena with a light covering of coarse vermiculite and then water the plug trays with half the amount of water that we would normally use for other crops. Verbena can even germinate without watering before putting into a good fog chamber for Stage I.

We conducted an experiment using four verbena varieties in which we either germinated them with or without a vermiculite cover and with or without a watering before putting them into a fog chamber. The highest germination occurred when verbena seeds were covered lightly with coarse vermiculite—either with or without a watering—before being put into a fog chamber.

Table 1. Plug germination percentage of different verbena varieties when germinated in a fog chamber at 78 degrees F.

| | Treatment | | | |
Variety	Uncovered & Dry	Uncovered & Watered	Covered & Dry	Covered & Watered
Showtime Blaze	59%	64%	75%	76%
Novalis Deep Blue	63%	56%	78%	76%
Romance White (Lot #1)	60%	64%	81%	84%
Romance White (Lot #2)	67%	64%	81%	83%

Experiments with four verbena varieties germinated with and without a vermiculite cover, and with or without a watering before being placed in a fog chamber. The highest germination was with seed covered with coarse vermiculite with or without a watering before being put into a fog chamber.

We tested other germination methods and combinations of watering and covering with coarse vermiculite, but none of them gave as high germination as sowing dry, covering with coarse vermiculite and either germinating dry or with a light watering before the fog chamber. Verbena needs this vermiculite covering for moisture control, not darkness.

In Stage II, continue to provide verbena with less moisture than you would normally give other crops. Reduce temperature to 72 to 75 degrees F. Provide light between 400 and 700 footcandles. A light feed of 50 ppm N from 20-10-20 can be applied every other watering once cotyledons expand. Keep medium pH between 5.8 and 6.5, with soluble salts less than 1.0 mmhos/cm^3 (2:1 extraction).

Many growers use a fungicide drench at time of sowing on verbena. We have not had to use any fungicides if we follow the reduced moisture method from time of sowing. Also, verbena are not heavy feeders, so keep fertilizations light and soluble salts low.

High-quality verbena seeds are now available as Genesis Seed. You can obtain higher, faster and more uniform germination using Genesis Seed combined with the above methods. Single-seeding now becomes a possibility with verbena.

Reduce water in Stage II for pansy success

Pansy seed has a wide range of temperatures for germination, anywhere from 62 to 78 degrees F. Reduced germination will occur if temperatures consistently go above 80 degrees F, as in summer and fall production.

Moisture levels in Stage I should be high (95 to 100 percent) until the radicle emerges from the seed. At that point, it is critical to reduce the moisture (75 to 80 percent) as with vinca. If pansies are kept too wet for too long, total germination will be reduced and will be very erratic.

We use a light covering of coarse vermiculite for moisture control. This covering is applied after seeding and watering for best germination results, as shown with vinca. Pansy seed does not need dark to germinate.

For Stage II, reduce temperature to 68 to 75 degrees F. Again, moisture levels must be reduced for good germination. Allow the medium to dry out slightly between waterings. Provide light between 400 and 700 footcandles. Keep medium pH between 5.5 and 5.8, and soluble salts less than 1.0 mmhos/cm^3 (2:1 extraction). Apply a light feed of 50 ppm N from 20-10-20 when cotyledons are fully expanded every other watering. Pansies, like verbena, are light feeders and do not like high soluble salts in the plug tray.

Genesis pansy seed has already demonstrated higher, faster and more uniform germination than standard seed in the industry. Growers must pay attention, though, to the upper temperature limit and the need for reduced moisture in Stage II to have success with pansy seed, whether using Genesis Seed or standard seed.

Fibrous begonias—wet and warm are key conditions

Warm and moist are the rules for good germination of fibrous begonia. Maintain soil temperature at 80 degrees F. If lower, germination may be reduced and will definitely be slower. Keep soil moisture high (95 to 100 percent) throughout Stage I and into Stage II. Fog chambers work best, but good germination can be achieved on the bench using fine mist or boom irrigators with a porous plastic covering such as Agricloth. Take care to make sure watering is done lightly after sowing in order to avoid burying the seed.

Increase germination in Stage I with the addition of 150 to 250 footcandles of light. Definitely in Stage II, begonias benefit from using HID lights providing 400 to 700 footcandles for at least 12 hours per day.

Once cotyledons have emerged (Stage II), reduce temperature to 75 degrees F, but continue to maintain high and even moisture levels. Begonia seedlings may stall if allowed to dry during this stage. Make an application of 50 ppm N from 20-10-20 every other watering during Stage II and early Stage III. Keep media pH between 5.8 and 6.5, with soluble salts less than 1.0 mmhos/cm^3 (2:1 extraction). Ammonium levels should be less than 15 ppm for best germination and seedling growth.

With these moist conditions through Stage II, algae growth becomes a big problem with begonia. The best control is with continuous use of Agribrom in the water supply at 10 to 15 ppm from the time of seeding.

Due to the small seed size, just sowing begonia accurately is a major task. Pelleted seed is available to make sowing easier. For best results with pelleted seed, maintain even and high moisture during Stage I. If the pellet is the type that melts after applying moisture, then apply a thorough watering after sowing. If the pellet is the type that cracks open, then keep from oversaturating the seed with the initial watering and maintain an even and high moisture level until the seedling emerges. If the pellets dry out at all during Stage I, germination will be reduced.

Fibrous begonias are also available as Genesis Seed. For best results with Genesis, follow the above procedures.

Attention to detail pays with these crops

Keep in mind that all four of these crops have specific requirements for good germination. Vinca likes it warm, moist and covered, but must have the moisture reduced once radicles emerge. Verbena likes it dry and covered. Pansies want high moisture with a cover, but are not as fussy about temperature unless above 80 F. Make sure to reduce the moisture for pansies after radicles come out, just like vinca.

Begonias demand warm and moist conditions during Stage I with some light. Keep the temperature and moisture up during Stage II, start light feedings and use HID lights. Genesis Seed will improve germination results with all four of these crops. **"**

Dr. Roger C. Styer and Shawn Laffe are in Plant Research and Development, Ball Seed Co., West Chicago, Illinois.

Water relations of seed germination: your key to successful plug growing

December 1989

by Paul Karlovich, Mehrassa Khademi and Dave Koranski

W ater is one of the driving forces of germination. Seed must attain and maintain an adequate water balance to germinate. Dry seed, due to their low moisture content, are relatively inactive. Seed, allowed to absorb moisture, become increasingly active and must be placed under the correct conditions to germinate properly or irreversible damage may occur.

Beyond this, plug producers do not want seed simply to germinate; they want them to germinate as quickly and uniformly as possible. Seed absorb water passively from surrounding medium in response to a water potential gradient, so the water potential of the seed must be sufficiently high so as not to hinder germination.

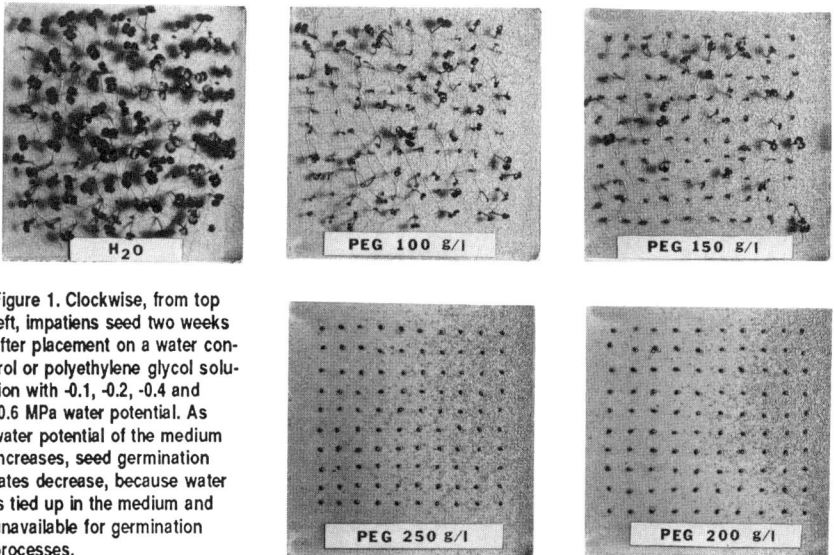

Figure 1. Clockwise, from top left, impatiens seed two weeks after placement on a water control or polyethylene glycol solution with -0.1, -0.2, -0.4 and -0.6 MPa water potential. As water potential of the medium increases, seed germination rates decrease, because water is tied up in the medium and unavailable for germination processes.

Defining water potential

The water potential is a measure of how "free" the water is. Pure water is totally free and has a defined water potential of zero. Adding fertilizers, sugars, fungicides, etc. to water makes that water less free, because some of the water molecules are attracted to the additions to the water. Thus the water potential becomes negative. In a similar manner, water in seed is less free than pure water because of water attraction to seed membranes, proteins and other seed structures.

When two systems are in contact with each other, water will always flow from the system with the freest water to the other system. Examples of such systems of concern to plug producers are the medium/seed, the medium/atmosphere and the seed/atmosphere systems. These three, two-system examples are the most important with respect to seed germination.

Dry seed absorbs water quickly

Dry seed have a **very** low water potential. Water potentials as low as minus 100MPa have been reported (compare this to the permanent wilting point for plants of minus 1.5MPa).This number indicates that water in a dry seed is not very free and that exposing seed to any system containing water is likely to cause the seed to absorb water.

For example, using your seeder in a damp location allows seed to absorb water from the air (the second system). If these seeds are then placed back 20in storage, they won't store as well as dryer seed. Studies with impatiens seed show that they absorb almost all the water necessary to germinate in three to four hours after being placed in a germination environment (Figure 2). Most flower seed absorb water in a similar time frame.

Imbibing seeds

Since a dry seed has such an extremely low water potential, a seed absorbs water very quickly when placed on even a moderately wet medium because of the great difference in water potential between the two systems. As water is absorbed by the seed, however, the water potentials equilibrate and water uptake slows down. At this point, germination may be disrupted if the seed cannot maintain an adequate water potential.

In research conducted at Iowa State University, a water potential as high as minus 0.1MPa (that corresponds approximately to an EC of 2.7) delayed impatiens germination (Figure 1). Lower water potentials delayed germination even further; minus 0.6MPa completely inhibited germination.

These water potentials do not affect initial water uptake in seed. Figure 2 shows a typical water uptake pattern (fresh weight accumulation) for many different seed types. Initially there is a rapid water uptake. The seed is so dry at this beginning point of germination that interference with water uptake probably doesn't occur.

Water uptake then stops or slowly increases. Prior to germination (radicle emergence), the seed spends most of its time at a certain moisture level. When the seed germinates, water uptake rapidly increases as the radicle elongates and cell division begins.

Three causes of water stress

Because dry seed absorbs water readily, most water-stress induced germination problems probably occur after the initial phase of water uptake. What can go wrong? Three potential causes of water-stress are dry media, high soluble salt and high evaporative demand. All three can produce the same results: slow and non-uniform germination and/or poor germination percentages.

Dry media results in desiccation

If the moisture level of the soil is too low, this will not allow enough water to get into the seed. A *lack* of water results in a low water potential, which can slow or stop water uptake and, in extreme cases, actually cause the seed to lose water to the medium.

Generally this is unintentional, but it happens when one is trying to keep conditions from being too wet. Occasionally drying of an entire crop occurs. More common is spotty drying caused by air drafts, temperature variations, light variations, different media levels in the plug cell, improper watering, etc.

This type of water stress happens more than most of us would like to admit and may significantly affect final saleable plug numbers. It is a serious problem because dry media results in severe desiccation of the seed or young seedling. This stress often kills the seed or seedling or so severely stunts growth that weak, slow growth results.

High soluble salts slow absorption

If the salt content of the soil is too high, water uptake by the seed can be slowed. Excess fertilizer and other salts reduce the water potential of the water in the medium. As explained above, salts in the water attract water molecules, and this water is not as free as if the salts were not present.

Salt stress is well known to any grower. It can be caused by high fertility programs combined with poor leaching practices, improper mixing of fertilizer additives to the medium resulting in locally high fertilizer concentrations, or drying the medium too far. Seeds that have not germinated can be kept from germinating or germinate erratically by high salt levels.

Salts lower the water po-

Figure 2.
Fresh Weight

A typical water uptake curve.

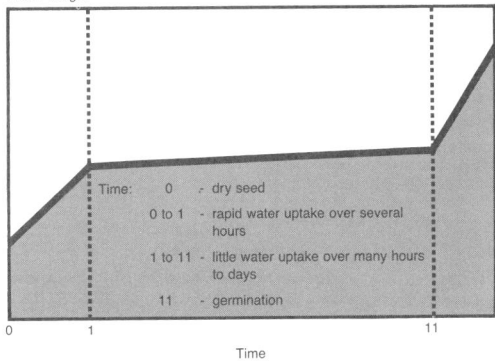

Time: 0   - dry seed
        0 to 1   - rapid water uptake over several hours
        1 to 11   - little water uptake over many hours to days
        11   - germination

Time

tential that, if too low, does not allow the seed to germinate. In fact, the controlled use of salts in water is used to prime seeds. Controlled salts levels can and are beneficial; uncontrolled salt levels are not.

Generally, high salt levels do not kill the seed, but they can have the same effect by delaying germination to the point of not having a saleable plug. The symptoms will be slow, erratic germination. High salt levels may be more damaging to young seedlings. Symptoms include root stunting and wilting. Young seedlings can be killed by high salt directly (burned roots) or indirectly by pathogens that attack weakened seedlings (Thielaviopsis on vinca, for example).

Dry air creates high evaporative demand

If the air atmosphere is too dry, water is lost more quickly from the seed than it can be taken up from the medium. The water potential of the atmosphere is very low, and water moves as vapor in response to the water potential gradient between the seed or the medium and the air. A seed can lose water as fast or faster as it takes it up from the medium. When this occurs, germination is inhibited or stopped, or the emerging radicle is stunted. This problem is magnified if the medium is too dry or if you have high soluble salts.

Stages 1 and 2 are most sensitive to desiccation of this form. As the seedling root system develops, a dryer atmosphere is not so great a problem because the roots can replace the water lost by transpiration.

Control moisture around seed

This discussion points to the need to control the moisture environment around germinating seed. With impatiens, we can say that any conditions subjecting the seed to a water potential of minus 0.1MPa (EC=2.7) or lower will delay germination. Avoiding such conditions is a must.

So what does minus 0.1MPa mean? If this water potential occurred in the soil medium during routine pot plant production, this would be considered a dry medium and would not be satisfactory for optimum growth. In short, medium moisture this dry should not occur often in potted plant production.

If this is so, does this mean that it is difficult to reach a minus 0.1MPa water potential during seed germination? No! And the reason we really need to be concerned is because the medium moisture level, the soluble salt concentration and the evaporative demand all interact. Flower seed are so small that they have very little tissue volume to buffer water loss against any changes in water potential.

We provided some basic information to explain why the proper moisture levels, particularly for Stages 1 and 2, are so important. Failure to do so is a major problem. Iowa State recommendations are to maintain a moist medium, a soluble salt concentration of less than 1.0 mmhos/cm³, and high relative humidity around the seed.

"

Paul Karlovich and Mehrassa Kahdemi are recent Ph.D graduates and Dave Koranski is professor of floriculture, Iowa State University, Ames.

Priming perennial seed

December 1988

by Dina M. Samfield, Jayne M. Zajicek and B. Greg Cobb

Texas A&M researchers increase Coreopsis lanceolata *germination from 47 percent to 81 percent using osmotic seed priming techniques.*

Plug production requires rapid, uniform germination and seedling emergence for high quality, uniformly sized plants. Osmotic seed priming techniques have improved rate and uniformity of seed germination and enhanced seedling vigor.

Priming coreopsis at Texas A&M

Recent research studies at Texas A&M University have shown that osmotic seed priming can significantly increase germination rate, percent and uniformity in a number of flowering annuals and perennials.

With *Coreopsis lanceolata* seed, priming for an average of 4.5 days in a 0.75 percent solution of potassium phosphate increases germination percentages. In this study, median germination time for primed coreopsis seed was 4.4 days, three days before that of untreated seed (7.4 days). Final percent germination of the primed seed (81 percent) was 34 percent higher than that of untreated seed (47 percent).

In the greenhouse, seed was planted at three times the recommended depth to provide an indication of primed seed vigor. At the end of four weeks, primed coreopsis seed had achieved 71 percent emergence, compared with only 56 percent emergence for untreated seed. At the same time, the percentage of seedlings at a minimum height of 1 centimeter was 30 percent for treated coreopsis seed and only 12 percent for untreated seed.

How does osmotic seed priming work?

When a solute such as salt is added to water, it reduces the amount of water in the solution, which in turn limits the amount of water taken up by seed. This prevents seed from germinating and extends the length of time during which major pregerminative metabolic processes occur.

When seed submerged in a salt solution such as this are surface dried and later sown, they again rapidly imbibe water and show accelerated root emergence. Since certain metabolic processes occurred during priming, they do not have to be repeated; germination occurs more rapidly.

Advantages of priming seed

For many crops, seed priming results in more rapid germination, increased germination percentages, improved germination uniformity and hastened rate of seedling emergence following sowing. In addition, seed with specific temperature requirements can be primed for successful germination over a wider range of temperatures.

In many instances, priming has been proven to promote seed and seedling vigor. A consistent improvement in time and synchronization of emergence achieved by osmotic seed priming results in improved stands, increased plant fresh weight and increased yield in number of crops.

Measuring the effects of seed priming

There are various methods being used to measure good quality seed and the effectiveness of seed priming.

Seed viability is measured in terms of the ability of seed to germinate. Viability is represented by the germination percentage.

Germination rate, prompt and uniform germination, is one of the more useful indicators of the effectiveness of priming. This is often expressed as median germination time, the time in which 50 percent of a seed lot germinates. The best priming agents enhance germination rate by reducing median germination time without decreasing total germination percentage.

Another measure of effectiveness of seed priming is vigorous seedling growth. Seed vigor is different from seed viability in that the latter is a yes or no response. Seed vigor is a quantitative response showing growing ability of ordinary viable and germinable seed under somewhat suboptimal conditions. Seed with low vigor may not be able to withstand unfavorable conditions in the seed bed, may be more susceptible to disease or may lack strength to emerge if planted too deeply.

Priming in combination with vacuum storage

An important objective of osmotic priming is to maintain improved seed vigor for the duration of seed storage. In a study to determine effects of vacuum storage on benefits of seed priming, primed and untreated coreopsis seed were placed in vacuum-packed cans and stored at 15 degrees C (59 degrees F) and 50 percent relative humidity for two months.

Identical treatments were repeated just prior to terminating vacuum storage, thus producing four treatments: an unstored control (untreated seed), stored control, primed unstored seed and primed stored seed. Seed from all treatments were sown at the same time in the lab and in the greenhouse.

Results from this study indicate that both vacuum storage and seed priming enhance seed germination of coreopsis. Of the non-stored seed, primed seed had the highest rate of germination, but it was stored, three-day primed seed that maintained the highest percent germination throughout the study.

In the greenhouse, seedlings of stored coreopsis seed averaged 14.6 percent higher total emergence than those of seed not stored. Seedlings of vacuum-stored seed appeared to be much larger and more uniform than those of seeds not vacuum-packed.

Additional work is now underway to examine methods of long-term storage that can preserve advantageous effects of seed priming. Researchers are also conducting studies to evaluate effects of specific salts, concentrations, osmotic potentials, priming temperatures and durations of priming for a broader range of plants. This information should be extremely useful to growers interested in priming their own seed for improved performance.

Dina M. Samfield is a graduate student and Jayne M. Zajicek and B. Greg Cobb are assistant professors with Texas A&M University, College Station.

GrowerTalks on Plugs

Priming coreopsis is not a complicated process

Priming is basically not a complicated process. The procedure and system outlined below is simplified and was worked under laboratory conditions with coreopsis seed.

Before committing any quantity of seed to osmotic priming, conduct preliminary tests with small quantities of each species. Each species responds differently to osmotic priming, some positively and some negatively. Some species may not respond to priming at all.

When experimenting with difficult-to-germinate seed, remember that some species cannot be primed and some seed species are sensitive to salts, and if the seed is treated, it will be destroyed. We have been using potassium phosphate and sodium chloride (table salt). Because salts may be detrimental to certain types of seed, preliminary testing on small seed quantities is required before committing large quantities of seed to osmotic priming treatments.

The optimum period of time for which seed is primed varies by species and variety and depends on the temperature at which they are treated. As a general rule, the warmer the temperature, the shorter the amount of time they need to be treated.

A 2-liter bottle of salt solution aerated by aquarium air pumps is typical priming apparatus. Seed are suspended in the salt solution for variable amounts of time, depending on species/variety and solution temperature.

Priming equipment

An understanding of the procedure and equipment we used may be helpful to growers working with difficult-to-germinate species such as many perennials. The equipment we used to prime in the lab is really quite simple:
- one 2-liter glass or plastic bottle or large jar
- 1/4-inch flexible plastic tubing
- one 115-volt aquarium air pump
- tape
- 1 liter of distilled water
- 18 grams (0.6 ounces) table salt
- a small strainer or aquarium net
- distilled water for rinsing seed following priming

Priming procedure

Pour 1 liter of distilled water into the 2-liter bottle. Add 18 grams of table salt and stir until completely dissolved. Place seed in a small square of cheesecloth along with two to three glass marbles. Carefully fold the cloth into a packet and secure it with string.

Suspend the cloth seed packet and the tubing connected to the aquarium air pump in the salt solution. Make sure the solution is bubbling and that seed is completely submerged. (To ensure adequate aeration, use not more than two, 2-liter bottles per aquarium pump.) After the system is adjusted, seal the top of the bottle with tape.

When working with difficult-to-germinate species, begin by testing several priming durations. Observe the point of time that seed begins to germinate in the solution. Reduce time accordingly, so that seed is activated to the brink of germination but does not actually germinate prior to sowing.

After priming, place seed in a strainer or net and rinse thoroughly with distilled water for at least one minute. Spread rinsed seed on dry paper towels and allow them to air dry until no longer damp (about 24 hours) before sowing.

Single-sown petunia, vinca and verbena flower faster in the flat

December 1988

by David S. Koranski, Julie A. Martens and Peter Lawlor

Plug trays often sport cells with several seedlings, resulting from a combination of events. Growers multiple-sow plugs to insure high germination rates for crops with low germination percentages. Seed size and seeder accuracy can also affect seeding density.

Past technology has not pursued long-term effects of seeding density. Current research, however, has shown that for several bedding plant crops multiple seeding does have an impact on plant growth, creating delays in flowering and decreased plant vigor.

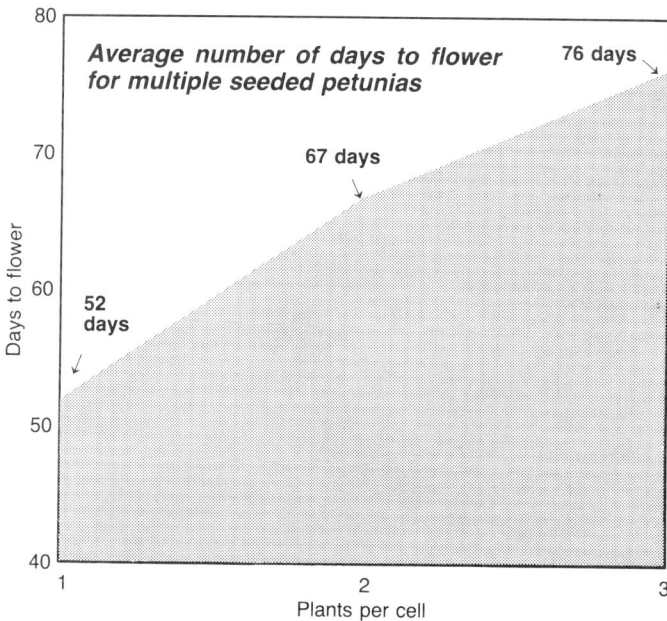

Average number of days to flower for multiple seeded petunias

Seeding density: Exploring an unfamiliar avenue

Looking at effects of seeding density, several researchers simultaneously conducted experiments following multiple seeded plugs from the plug tray to the finished product.

In trials at Iowa State, Dr. David Koranski sowed Red Madness petunia seeds in 406 standard plug trays using a seeding density of one, two or three seed per cell. Following germination, he added seedlings to cells lacking the required number of plants.

Dr. Roger Styer of the Ball Seed plant research and development department conducted experiments in West Chicago, Illinois, using five annuals: Ultra Pink petunias, Prelude Rose begonias, Red Hot Sally salvia, Little Bright Eyes vinca and Showtime Blaze verbena. He sowed plug trays using a seeding density of one or two seeds per cell.

Multiple sowing delays finishing time

For all annual species used in the trials, plant growth declines as the number of seeds per plug tray cell increases. Delays in flowering occur as seeding density rises from one to two to three seedlings per cell.

Petunias show the most dramatic response to seeding density, requiring one to two additional weeks to bloom when double-planted and three more weeks when triple-planted. Plant growth decreases with each additional seedling per plug cell, as the plants compete for nutrients and growing space. For the double-seeded plants, a two-week delay results between transplanting and the time to saleable flat. Although plant growth declines for all other annuals as seeding density doubles, times to saleable flats vary. For salvia and begonia, multiple sowing has no observed effect on time to saleable flat. Vinca and verbena show increases of seven and four days, respectively, to saleable flat stage with the double cell density.

The bottom line on density

As in all businesses, growers need to examine the impact of new technology on their profit margins. In the final analysis, seeding density affects plant quality and development of plug-grown annuals.

• Two or more seed per cell causes competition between seedlings, resulting in a decrease in both shoot and root tissue. This produces seedlings that, ultimately, are less vigorous.

• In some species, flowering may be delayed when plugs are multiple sown. Delays vary from 14 to 21 days in petunia to seven days in vinca to four days in verbena.

• One seed per cell not only results in a more uniform stand of seedlings, but also eliminates flowering delays for some species.

For some annuals, multiple seed sowing results in growth declines and flowering delays, which could mean the difference between matching or falling short of a competitor's crop's quality, meeting or missing a market deadline, and receiving or relinquishing top prices in the floral marketplace.

"

Dr. David S. Koranski is professor of floriculture, Iowa State University, Ames, Iowa. Peter Lawlor is a technician at Iowa State University. Julie A. Martens is managing editor of GrowerTalks magazine. The authors acknowledge Dr. Roger Styer for his work on this project and for his editorial assistance.

Don't bury your seed alive!

December 1988

by Paul Karlovich and David S. Koranski

Seed requires oxygen to germinate. Germination occurs most rapidly at oxygen concentrations approaching that of air, which contains 21 percent oxygen. While many seeds sown in plug flats remain uncovered, some crops are sown and covered with a shallow layer of growing medium.

Does shallow burial of seed in a plug flat affect germination? The real question here is whether or not seed experiences the oxygen level of air or whether the water layer that forms over seed subjects seed to some unkown, lower oxygen concentration.

Water and media can suffocate seed

To obtain an answer to the oxygen level question, Paul Karlovich, Iowa State University, exposed seed of Super Elfin Orchid impatiens to reduced oxygen concentrations.

He used a closed system, sowing seeds on blotter paper in sealed jars. Paul tested oxygen levels of 20, 7, 5, 3, 1 and zero percent. Oxygen levels of 7 percent or less significantly reduced germination rate.

At zero percent oxygen, the Super Elfin Orchid seed did not germinate in seven days at 77 degrees F. Surprisingly, Super Elfin Orchid germinated under water (with less than 0.01 percent oxygen) as fast as it did in 20 percent oxygen.

The fact that Super Elfin Orchid seed germinated so well under water caused Paul to doubt the significance of the results obtained with reduced oxygen concentrations.

If an impatiens seed germinates under water, then conceivably it should have no trouble germinating under any conditions in the plug flat, provided all other cultural conditions are correct and adequate moisture is provided.

Defining a relationship between germination and seed placement

Using a soilless mix (30 percent sand-finish perlite, 35 percent hypnum peat and 35 percent sphagnum peat) Paul sowed Super Elfin Orchid and Rose Star impatiens into 392 plug trays. He saturated the medium from below (subirrigated) for 24 hours with water and then placed seed on top of the medium or pushed seed 2 to 3 millimeters (one-eighth inch) into the saturated medium and covered.

Paul created three levels of moisture in the plug tray medium: (1) draining the tray on blotter paper for 30 minutes; (2) draining the tray to container capacity; and (3) keeping the tray wetter than container capacity by setting it in 5 to 6 millimeters (one-quarter inch) of water.

To prevent evaporation and to maintain high relative humidity, Paul covered the trays with plastic wrap. Low lights illuminated the plug trays in a growth chamber held at approximately 79 degrees F.

The emergence of the first root from the seed coat marked germination, which was measured after seven days. Paul counted germinated seeds placed on top of the medium and seedling emergence on the buried seeds. Buried seeds that failed to emerge were dug from the medium.

Paul chose to study Rose Star impatiens because, unlike Super Elfin Orchid, in germination tests under water Rose Star germination was much lower than in air. Also, Rose Star's final germination rate after seven days was lower under water compared to air.

After seven days it was clear that differences existed between buried seed and seed placed on top of the medium. For seed on top of the medium, Rose Star averaged 98 percent germination across all medium moisture levels; Super Elfin Orchid averaged 92 percent.

For buried seed, only 13 percent of Rose Star seed emerged from the medium, while Super Elfin Orchid had 35 percent emergence. When Paul dug up seed that had not emerged from the soil, he counted and added those values to the seed that had emerged. Final germination percentages for buried seed were 42 for Rose Star and 81 for Super Elfin Orchid. Paul is continuing his investigation by looking at additional impatiens cultivars.

Burying seed could stop impatiens germination

These results suggest that if an impatiens seed is buried in a plug tray, even to the depth of only one-eighth inch, the chances of it ever becoming a useable plug are extremely low.

Improper handling of plug flats after seeding could cause the seed to be buried, and if a significant number of seeds are buried, then erratic and slow germination may be a likely result. Even a small percentage of buried seed may reduce the final number of saleable plugs.

Media moisture is less crucial than oxygen level

Paul is specifically investigating the effects of Stage I conditions on impatiens' germination. After seven days, the growth of buried seed was visibly behind that of non-buried seed, even if they had emerged. Impatiens seed on top of the medium were not adversely affected by the medium moisture level.

The data presented above was the average of the three moisture levels tested. There were additional differences in germination and emergence among the moisture levels, but they were relatively small when compared with the differences seen between buried and non-buried seed.

Interestingly, seed of both impatiens cultivars germinate at higher percentages under water than when buried. Paul feels that there are three reasons for the poor germination and emergence of buried seed:

- low oxygen concentration.
- lack of light.
- mechanical resistance to emergence because of media on top of the seed.

With the seed lots tested in this research, lack of light could be a problem with Super Elfin Orchid impatiens, but it was probably not with Rose Star. At this point, Paul believes that a low oxygen concentration is responsible for the substantial reduction in germination percent and rate.

Handle plug flats gently to avoid burying seed

Improper handling of plug flats after seeding can cause seed to be buried. Careful attention should be paid to flat handling after seeding. Growers should question their plug-tray handling practices.

- What is happening to **your** seed?
- Are you bouncing seed into the medium while transporting it from the seeder to the germination area?
- Do your watering practices wash seeds into the medium, or worse, out of the tray?
- Is your medium tight enough to prevent seed from falling into large crevices in the medium?

Preliminary studies indicate that buried seed is a problem with impatiens. Other seeds such as petunia and begonia have not been studied to date, but until proven otherwise, care should be exercised throughout all of seed practices to maximize the amount of oxygen available for germination.

"

Paul Karlovich is a Ph.D. candidate at Iowa State University, Ames, Iowa. Dr. David S. Koranski is professor of floriculture, Iowa State University.

Shorten soak times and boost temperatures with primula

December 1988

by Tom Bruening and David S. Koranski

Temperature is the key factor affecting primula germination rates. Previously, growers practiced seed soaking to boost germination. Current recommendations for germination urge temperatures from 59 to 62 F, when in fact higher temperatures improve germination.

Primula: An off-season crop for all seasons

Primula popularity has increased worldwide. Breeders have improved the plant's color, shade and growth habit. Growers find primula to be advantageous for many reasons and are raising this attractive potted plant for off-season sales.

Plants are started in mid-June or early July, usually after the spring season is over. By that time, many benches have been cleared of bedding plants, work loads have begun to diminish and little supplemental heat is required to start this cool crop.

The plant flourishes in the cooler fall environment, and the compact growth habit allows good bench densities. In the spring, primula provides opportunities to market an easy-selling, long-lasting plant for late-winter sales.

Discovering how to overcome low germination percentages

Germination has presented consistent problems for growers. Germination rates as low as 20 to 50 percent are not uncommon, a fact that has made this otherwise popular plant less attractive to growers.

To make primulas economically viable, growers need adequate germination rates. Using gibberellins to enhance germination has limited success.

Tom Bruening, Iowa State University, conducted experiments to determine the effects of temperature and water soaking on primula seed germination. His objective was to improve germination rates with environmental conditions that could be easily controlled by growers.

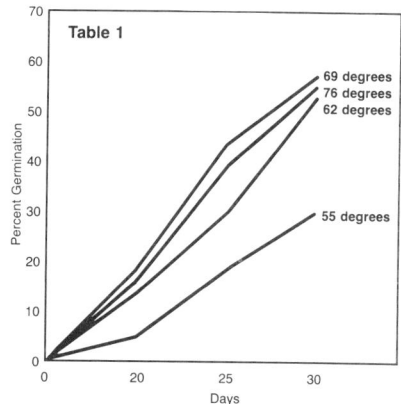

Table 1

A uniform lot of Yellow Pageant primula seed were soaked at 40 F for zero, two, four and eight days. Tom sowed the seed on two sheets of brown paper and seven sheets of white Kompac that retained the 20 milliliters of distilled water in 4-inch by 4-inch clear plastic petri dishes.

Each petri dish contained 100 seed and was placed in controlled environmental growth rooms maintained at 55, 62, 69 and 76 F. Moisture was added to maintain approximately 85 percent humidity; light intensity in the growth chambers was 500 footcandles.

Germination rates were recorded daily from the day seed began to germinate until day 30. Seed was considered germinated when the seed leaves were completely expanded.

Heating to 69 F boosts germination

For the temperatures used, there was a significant difference between the germination percentages. Primulas grown at 62, 69 and 76 F had germination that was greater than at 59 F.

For this variety and seed lot grown under the experimental conditions, the 69 F temperature appears optimum; the 55 F treatment provided poorest germination percentages. No significant differences existed between germination rates for 62, 69 and 76 F.

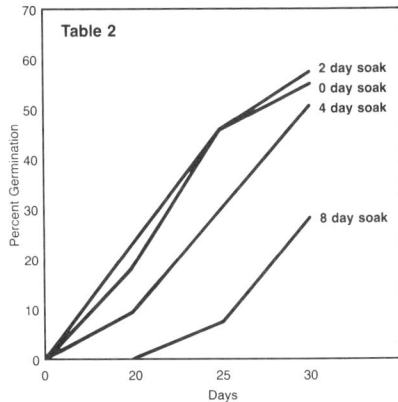

Table 2

A two- or eight-day soak? Shorter is better

Soaking primula seed resulted in a significant difference in germination percentages. Optimum soaking period was two days. There was little difference between final germination percentages at zero, two and four days.

Compared to eight days, however, there was a large difference in percent germinated seed for soaking treatments of zero, two and four days. At germination day 30, 52.5 percent of two-day-soaked seed germinated, while only 27.4 percent of eight-day-soaked seed germinated.

Primula seed soaked two days and grown at 69 F showed the best germination rates, with 62 percent of the seed germinating under this combination. With two days of soaking, 57 percent of seed germinated at 62 F, while 54 percent germinated at 69 F.

There was a difference between average number of diseased seed and seedlings when compared to number of days soaked. No soaking had 7 percent diseased seed; two soak days had 6 percent; four days had 5 percent. Eight-day soaking reduced disease loss to 4 percent.

How to enhance primula germination

Recommended temperatures for primula germination are usually given between 59 and 62 F. On the basis of this research, germination temperatures for Yellow Pageant primula should be increased. Optimum temperature may actually depend on seed age and maturity.

Because of the high price of seed, growers are urged to experiment with temperature in small lots well before planting time to determine optimum environmental conditions for germination at their locations.

The seed coat on primula is very hard. Soaking the seed prior to planting delayed germination in all treatments. Slight improvements in seed germination resulted with a two-day soak treatment, but the results were not significant.

Soaking four days or longer is not advised. In this research, prolonged soaking produced poorest germination percentages. Saturating seed with water for more than two days may inhibit germination.

99

Tom Bruening is a Ph.D. candidate at Iowa State University, Ames, Iowa. Dr. David Koranski is a floriculture professor at Iowa State University.

Primed Seed
A step beyond refined seed

January 1988

by David S. Koranski

When we first got into plug production, there were a lot of unanswered questions. We were encountering problems with germination and with growth and development of the seedlings. Was it a problem with the environment, or with the seed? We learned the difficulties were from both. Sometimes the seed product had less vigor than it should have, but the seed was being grown under less than optimum environmental conditions.

Throughout the evolution of plug production, we've worked out many of the requirements that are necessary to provide optimum quality plugs with the environment. At Iowa State a lot of work has been done to determine what the optimum environmental conditions are to germinate seed. Optimum temperature, light and moisture conditions have been established through Stage I and Stage II.

At the same time, the seed companies have done an excellent job in improving the vigor of seed. Refined seed represents some of the improvements. Seed is cleaned and then by gravity and density and weight the best, most uniform seed is separated and termed refined. But even with refined seed difficulties remained: The germination rate still covered a wide window and we still didn't have a product that germinated at a variety of temperatures. Many of our seed crops such as vinca, salvia, pansies and petunias germinate over a time period as short as three days and as long as 15 days. Growers have to provide optimum environmental conditions throughout the whole germination sequence because of uniform germination—up to 15 days. That's very difficult for growers to do.

Because of growers' need to have uniform germination, companies developed a seed product called primed seed (trade name Genesis from the Ball Seed Co., West Chicago, Illinois). Part of the process to create primed seed involves setting up an osmotic pressure to get water into the seed, thus starting germination. Then, the process is stopped or inhibited. However, throughout the whole process, the seed coat is not broken.

Our research group tested primed seed vs. nonprimed, or standard seed, to see if there was any difference in germination rates or percentages. The seed varieties we tested include Universal Blue pansies and Super Elfin Orange impatiens.

		Day 4	Day 5	Day 6	Day 7	Day 8	Day 9	Day 10	Day 11	Day 12	Day 13	Day 14
							% germination					
Temperature												
59°F	P	—	—	—	12.5	40	57	59	84	92	91	91
	NP	—	—	—	—	—	—	—	6.5	42	55	54
66°F	P	—	46	62	69	79	85.5	84	90	94	94	94
	NP	—	—	—	1	4.5	23	26	61.5	75	76	76
73°F	P	—	47.5	75	76	82.5	84.5	85.5	89	91	92	92
	NP	—	—	2	5	11.5	27	35	47	62	68	68
80°F	P	—	1.5	19	37	68	79.5	80	85.5	90	89.5	89.5
	NP	—	—	—	5	15	41.5	43	63	77.5	78.5	79

Table 1: Comparison of primed and nonprimed seed germination percentages of *Viola × wittrockiana* Universal Blue with four germination temperatures. P = Primed seed   NP = Nonprimed seed

	Day 4		Day 5		Day 6		Day 7		Day 8		Day 9		Day 10		Day 13	
	P	NP	P	NP	P	NP	P	NP	P	NP	P	NP	P	NP	P	NP
								% germination								
63°F	—	—	—	—	—	—	—	—	—	—	48%	28%	98%	71%	99%	91%
70°F	—	—	—	—	92%	77%	98%	97%	99%	98%	99%	98%	99%	98%	99%	98%
77°F	33%	25%	82%	70%	98%	98%	99%	99%	99%	99%	99%	99%	99%	99%	99%	99%
84°F	67%	50%	96%	93%	99%	97%	99%	98%	99%	99%	99%	99%	99%	99%	99%	99%

Table 2: Comparison of primed and nonprimed seed germination percentages of *Impatiens wallerana* Super Elfin Orange with four germination temperatures. P = Primed seed   NP = Nonprimed seed

Materials and methods: Seed was sown into 406 plug trays using a medium consisting of 65 percent coarse grade spagnum peat moss, 15 percent vermiculite, 15 percent sand finished perlite and 5 percent calcined clay. The medium was amended with 2.97 kilograms per cubic meter calcium carbonate, 1.78 kilograms per cubic meter of minor elements consisting of 1 percent sulfur, .02 percent boron, .3 percent copper, 2.0 percent iron, .5 percent manganese, .006 percent molybdenum, 1 percent zinc and not more than 2.6 percent chloride. Calcium carbonate and single superphosphate were ground before incorporating them into the medium. After the 406 plug trays were filled, they were placed overnight under intermittent mist for moistening before sowing. Each flat, representing Universal Blue or Super Elfin Orange impatiens was divided into four sections each containing 100 cells.

After sowing, flats were placed in a growth chamber at temperatures of 59 F, 66 F, 73 F or 80 F. Moisture was provided with fine mist of about 35 microns. Humidity was maintained at approximately 95 percent and light intensity was approximately 500 footcandles. Germination rate was recorded daily from the day seedlings started to germinate until day 14. Seed was considered germinated when cotyledons were expanded completely. At 15 days after sowing, the number of no-sheds and abnormals was recorded. No-sheds were those seedlings unable to shed their seed coats, while abnormals were translucent seedlings that grew horizontally on the soil surface.

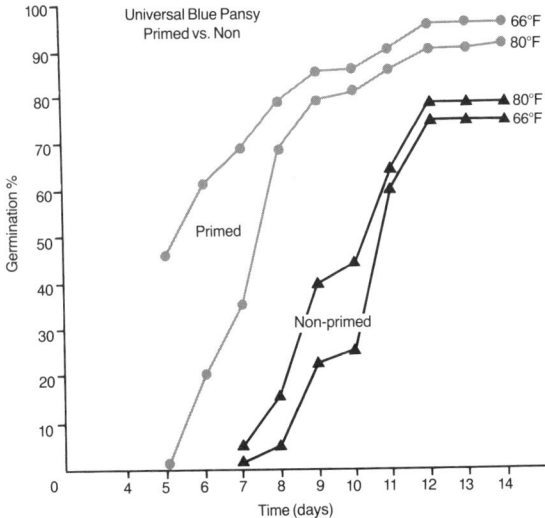

The experiment was repeated three times.

Results: Universal Blue pansies

There was a major difference in the germination rate of Universal Blue pansies between primed seed and non-primed seed as is demonstrated in Table 1. The table shows that there is a very fast germination rate and a very high germination percentage for primed seed at all given temperatures. With the nonprimed seed, there is a slow germination rate and a reduced germination percentage at any given temperature.

The optimum temperature for the germination of primed seed was 66 F. The optimum temperature for nonprimed seed was 80 F. With the primed seed, there was very little difference in the germination percentages at day 14 at any temperature. However, there was as much as 54 percent total difference in germination at 59 F for nonprimed seed and 89 percent difference in germination at 80 F. Another interesting point is that if you look at the beginning of germination from primed seed, at 66 F, 69 percent of the seed germinated on day 7 and at 73 F, 76 percent was germinated. For nonprimed seed at the same temperatures, germination percentages were 1 percent and 5 percent respectively. Primed seed germinated best at 66 F and the nonprimed seed germinated best at 80 F; however, at any given temperature, the primed seed germinated uniformly, and nonprimed seed showed a vast difference in germination percentages.

Super Elfin Orange impatiens

Based on the information in Table 2, it is obvious that germination percentages at day 13 are extremely good for both primed and unprimed Super Elfin Orange seed. If you look at temperature, you still get 99 percent germination for primed seed at 63 F, 70 F, 77 F or 84 F.The rate of germination, however is linearly affected by temperature: As temperature increases, the rate of germination increases, which is not startling. But with primed seed, at 84 F you get 76 percent germination at day 4, which is extremely high for impatiens. There is a difference between primed and nonprimed impatiens germination rates, but it is not as great as the differences exhibited in primed vs. nonprimed pansy seed. The most important point is that germination rate is faster in primed impatiens seed. Germination percentages are almost identical, but the rate of germination is faster, primed seed peaked at day six.

What does this mean to the grower?

Primed seed is faster and more uniform. It will give the grower a much higher germination percentage for pansies and a much faster germination rate for impatiens. The reason for the performance is that primed seed has already started to germinate internally. When it is put under optimum germination conditions, it just germinates much faster.

The implications of these results for the industry and plug production are multi-fold. With primed seed the grower can germinate at higher percentages regardless of temperature. The germination rate is very fast in that most of the seed is germinated in the first five days.

Growers now can grow seed at any temperature and may not have to have optimum environmental conditions as is needed for standard or refined seed. More growers might be able to grow plugs with primed seed.

Primed seed is also very uniform and fast; therefore the time in the germination chamber is reduced and time in germination facilities is much less. Since there isn't as much time in the chamber or on the bench, cost of production is less.

The bottom line is that you can germinate primed seed at nearly any temperature with good germination percentages. Primed seed is much faster, so the investment in facilities to produce plugs are not as great.

"

David S. Koranski is professor of floriculture, Iowa State University, Ames.

A look at refined seed

January 1987

by Ann Reilly

It's known by a lot of different names, including High-G, High Tech, Fresh Tech, High Vi, Super Seed, or High Energy. The seed industry prefers the generic terminology "refined seed." There's a lot of things that can be said about it, but there are still a lot of unknowns, mostly because a great deal of the information is privileged within each of the seed companies.

By way of definition, it would be fair to say that refined seed is a seed of predictably high germination, with high vigor, high energy, and with a uniform germination and growth rate. It has been around the industry for only a few years and is the subject of promotion, discussion and disagreement among seed producers and bedding plant growers alike. It was developed for the plug industry where high germination and uniformity of germination and growth are critical.

Some seed companies use it as a promotion tool. With others, you're hard pressed to find mention of it within their catalog. One admitted that the only reason they carry it is because their competition forced them into it. It is more expensive, and the price differential varies greatly from seed producer to seed producer. Some growers think the increased price is worth it. Others don't.

It is more expensive

Vaughan's Seed Company, a seed broker, buys some seed that is already refined from the seed producers. When this happens, they do nothing further to change it; they rename it Super Seed and pass it along to their customers. The availability list includes begonias, dahlias, dusty Millers, impatiens, petunias, salvias, detailed marigolds, and defuzzed tomatoes, in limited varieties. Super Seed is listed in the catalog, but is a far cry from a main promotional feature. Prices run at least 10 percent higher, and total volume of super Seed sold is about 2 percent of sales. As long as the growers want it, perceiving it to be worth the extra cost, Vaughan's will sell it. They see the entire program as a product of grower perception driven by the seed producer.

At Vaughan's neighbor, Ball Seed Company, there is more promotional effort devoted to their High Energy Seed. They offer more than 100 varieties of ageratums, begonias, coleus, impatiens, pansies, petunias, salvias, vincas, detailed marigolds, and defuzzed tomatoes, and state that germination is rapid, uniform, and at least 90 percent. This germination guarantee also varies from seed company to seed company; Park Seed, for example, guarantees 95 percent. However, regular seed is subject to discounts depending on order size, so the price could be reduced to the point that the premium for High Energy Seed could be anywhere up to 25 percent. This is in line with other seed companies that report anywhere from a 10 percent to 40 percent premium for refined seed, depending on the type.

Exactly what is refined seed?

What is the process that is used to achieve refined seed? While no one would say exactly, Goldsmith revealed that they refine petunias based on seed density and impatiens based on seed size. Park Seed Company explained it as a combination of specific gravity and/or surface tension and/or size, depending on the type of seed, with final selections going through a 2-step process. Harris

It's known by a lot of different trade names, including High-G, High Tech, Fresh Tech, High Vi, Super Seed or High Energy. The seed industry prefers the generic terminology "refined seed." Above, raw ageratum seed. Right, Harris' High Tech seed. The High Tech seed has been detailed and "super cleaned" to remove dirt and other inert material.

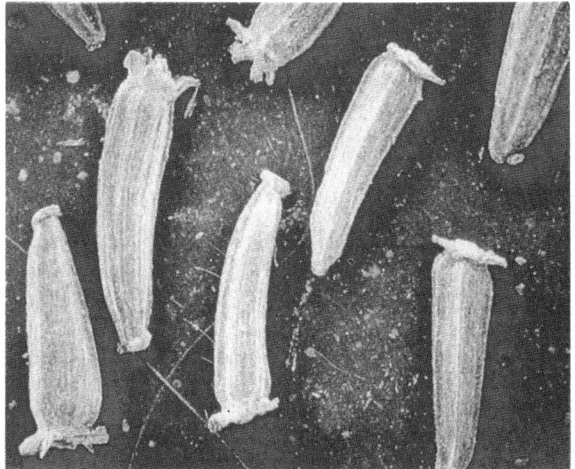

Moran describes it as a combination of mechanical and chemical treatments. A new scarification method for geraniums is refining all geraniums. The bottom line is increased uniformity and vigor. Some years many varieties may be available as refined seed; other years, only limited varieties will be available. The industry hasn't yet learned to refine some types.

One thing all seed companies do agree upon is that they don't want the customer left with the impression that "unrefined" seed is the leftover, the junk, the portion that should be trashed. On the contrary, regular seed is of as high quality as it ever was. The difference is that it may not be of the same uniformity or vigor, but it still will produce a crop just as acceptable, although perhaps not as quickly.

The refined seed program will actually have the effect that, one day in the foreseeable future (about five years), all seed sold will be of the quality of today's refined seed. The list of available refined seed is growing. Park Seed, for example, offered the first refined seed four years ago. That seed was quite expensive, perhaps as much as double in price. There were 30 varieties the first year, and now there are 150 or more at Park, including ageratums, begonias, carnations, gerberas, impatiens, lisianthus, pansies, petunias, salvias, snapdragons, and eight vegetables.

As we have already said, one day all seed will be refined seed. Park goes a little beyond that in sophistication and talks of seed that may be refined for temperature and climate differences; for example, pansies for germination at 55 F and pansies for germination at 70 F. This would be especially helpful when dealing with geographic or timing differences, such as sowing pansies in Texas in summer for fall sales. There's also talk of pre-soaking seed to increase germination, a process made possible by causing water to enter the seed through differences in osmotic pressure. This has been done successfully with turfgrass seed such as Kentucky bluegrass varieties, which are slow to germinate.

At Northrup King, they've taken the refined seed, which they call Fresh Tech, and combined it with a resealable zip-lock package named Fresh Pack. Being marketers, not producers, they are trying to zero in on the market, identify the specialty growers and provide them with a quality product that will give them a greater return per square foot. They know the seed is high quality, so they combined it with the new package to increase their market share.

Harris-Moran, which was one of the first seed companies to offer refined seed, admits it was one of the best things they have done to increase their market share. They describe it as a "marketing tool with value." They also agree that one day all seed will be refined seed, if seed has not become obsolete before that time comes. That's a whole other topic but an intriguing one, no?

Is it the answer for everyone?

So, should all growers turn to refined seed where available? Not necessarily. A smaller grower will most likely find it not economically feasible. The larger grower, especially the plug grower, is the major market for refined seed. It is easy to use with mechanical sowers, and with its high germination, it leaves few blanks. This naturally relates to a savings in money. Uniformity in germination and growth produces a flat where all seedlings are ready for transplanting or sale at the same time, reducing handling. So, if you are a large plug grower, the increased price is probably worth it. But there are no guarantees. You'll have to try it yourself to be sure. You can be sure of one thing if you fit the prime user category. Refined seed is definitely worth the try.

"

Ann Reilly is a freelance writer residing in Belize, Central America.

Media, Fertilizer and Water

Too B or not too B

December 1989

by Shawn R. Laffe and Roger C. Styer

Learn to recognize boron deficiency.

Many common greenhouse problems come and go in waves, and so too the problem of boron deficiency. Reports throughout the country have identified boron deficiency in crops such as pansies, petunias and impatiens—in plugs, as well as finished containers. Times of intense production coincide with the greatest occurrence. Fall pansies and petunias, from California to Florida, and petunias produced in the late spring in the North have been the most susceptible crops in the past few years.

Symptoms of boron deficiency include abortion of growing tips, proliferation of axillary shoots, strapped or crinkled leaves, upward cupping of leaves, stunted growth and a slight chlorosis of upper leaves. The deficiency resembles an excessive Bonzi treatment. Bonzi, however, causes leaves to darken and won't destroy the terminal growing point or cause shoot proliferation.

Boron deficiency also resembles thrips or mite damage because of cupped and strapped leaves. Close inspection proves the presence or absence of any insect feeding scars (tiny yellow spots) or webbing on leaves.

Manage Ca, Mg and pH

Cultural practices are mainly responsible for boron deficiency. The problem occurs most frequently during hot weather. Growers water more often, thus leaching nutrients—including boron—from the soil.

Also in hot weather, growers often reduce fertilizer levels to prevent plant stretch. Applied fertilizer often doesn't replenish lost boron. High pH levels (greater than 6.5) may tie up boron in the medium, making it unavailable to plants. Boron becomes readily available at pH below 6.0 and especially at pH 4.5 to 5.5. Hold soil pH at 5.5 to 6.5 to ensure availability of most nutrients.

High calcium and low magnesium levels can also tie up boron. The balance of these nutrients one to two weeks prior to periods of stress may be a critical link to boron deficiency. Incorporate magnesium sulfate (Epsom salts) in the media or feed to bring magnesium levels in balance with calcium levels.

Avoid too much of a good thing

At operations where boron deficiency recurs year after year, try incorporating a dry fertilizer containing boron, such as Esmigran, into soil prior to planting. This makes boron available to seedlings in early growth stages. Applications of liquid peat-lite fertilizers, such as 20-10-20, can supply minor elements, including boron. At low parts per million, though, quantities may not be adequate for growth.

One or two applications of STEM (soluble trace elements) at one-half to full rate may be beneficial. Solubor at 0.25 ounces per 100 gallons or Borax at 0.5 ounces per 100 gallons can also supply boron. During periods of heavy watering and leaching, boron levels must be maintained. Plants severely affected by boron deficiency may not recover even after receiving application of a boron supplement.

Soil and foliar analyses, combined with early boron supplements and close observation, are tools growers can use to avoid or discover boron problems in early stages. According to soil test recommendation by W.R. Grace & Co., satisfactory soil boron levels range from 0.05 to 0.5 ppm. Levels less than 0.25 ppm have been associated with boron deficiency levels; in excess of 0.45 ppm, toxicity. In the foliage, boron should range from 30 to 150 ppm; in the water, less than 0.5 ppm. Water boron levels of 0.25 ppm and higher may be toxic to boron-sensitive crops.

Moderation is important. In an effort to correct or prevent boron deficiency, growers may apply "too much of a good thing," resulting in high boron levels. Boron toxicity appears as chlorosis and burn on margins of older leaves and stunted shoot and root growth. Sensitive crops include begonias, gerberas, marigolds, lantana, pansies and zinnias.

"

Shawn Laffe and Dr. Roger Styer are in Plant Research and Development at Ball Seed Co., West Chicago, Illinois.

Production 101:Sorting the relationship between water quality, feeding programs and media components

September 1989

by David S. Koranski

Water quality, media and nutrition interact to control your crop—from root development through flowering. The more you understand these basic building blocks of production, the better—and easier—you can grow. Here's the scoop in simple terms, from plugs to pH to finished poinsettias.

How should a grower fertilize a plant? One greenhouse grower wanted to incorporate phosphorus, magnesium and calcium in lily soil to produce wide leaves. He worked with soil, water and tissue samples. But it soon became apparent that, before he could choose his medium or select his fertilizer, the grower had to understand water quality.

Defining water quality

In general, there are several chemical properties of water that may cause problems for plant producers: pH, alkalinity, soluble salts, boron, fluoride, chloride, sulfate, sodium and iron.

● **pH of water** refers to a measure of whether the water is acid or alkaline. A pH of 7.0 is neutral, below 7.0 is acid and above 7.0 is alkaline. Since pH is measured logarithmically, a pH of 8.0 is 10 times more alkaline than a pH of 7.0. Keep water pH between 5.0 and 6.5—most elements and other chemicals, such as growth regulators and fungicides, are available at that pH.

● **Alkalinity of water** measures amounts of calcium and magnesium carbonates present. Alkalinity affects the ease or difficulty of reducing water pH; the

Table 1. Desirable ranges for specific elements in irrigation water.*

Phosphorus (P)	.005 to 5 mg/1	Zinc (Zn)	1 to 5 mg/1
Potassium (K)	0.5 to 10 mg/1	Sodium (Na)	0 to 50 mg/1
Calcium (Ca)	40 to 120 mg/1	Aluminum (Al)	0 to 5.0 mg/1
Magnesium (Mg)	6 to 24 mg/1	Molybdenum (Mo)	0 to 0.02 mg/1
Manganese (Mn)	0.5 to 2 mg/1	Chloride (Cl)	0 to 140 mg/1
Iron (Fe)	2 to 5 mg/1	Fluoride (F)	0 to 1.0 mg/1
Boron (B)	0.2 to 0.8 mg/1	Nitrate (NO_3)	0 to 5 mg/1
Copper (Cu)	0 to 0.2 mg/1	Ammonium (NH_4)	Undetermined
Sulfates (SO_4)	24 to 240 mg/1	Alkalinity	1 to 100 mg/1 $CaCO_3$
Soluble salts	0 to 1.5 mmhos	SAR^z	0 to 4
		pH	5 to 7

Gerald M. Curtice and Dr. A. Robert Templeton. Water Quality Reference Guide for Horticulture. Aquatrols Corporation of America, Inc., Pennsaken, NJ (December 1987).

higher the alkalinity, the more difficult it is to lower the pH. Water is said to be buffered or highly buffered, depending on the alkalinity reading. Alkalinity for plugs should be approximately 60 ppm; for pot plants, 100 ppm.

Influence of pH drift, caused by high alkalinity, is especially important with respect to micronutrient availability. Compensate for high pH and alkalinity levels by injecting acid—phosphoric, sulfuric or nitric—into irrigation water. Inject acid into irrigation water before fertilizers are injected.

• **Total soluble salts** concentration in irrigation water is made up of all the dissolved chemical elements in solution—your fertilizer, any acid additions, salts already present in your water and fungicides. High soluble salts concentration may increase soluble salt levels in a growing medium. Salt sensitivity is based not only on plant species being grown, but also on individual elements making up the solution. Numerous crops grown in plugs can be sensitive to high soluble salt levels in the medium. Try managing high soluble salt levels in plugs by using a coarse, well-drained medium and applying 5 percent to 10 percent excess irrigation water to facilitate leaching.

• **Boron, fluoride, chloride, sulfates** and sodium, if present in irrigation water at elevated levels, not only reduce plant quality but also affect soluble salt levels in water and growing medium. In some situations, growing at higher pH levels (6.5 to 6.8) manages elevated boron and fluoride levels by reducing their solubilities and making them unavailable to plants.

Test your water to determine types and levels of chemicals present. Modify an existing fertility program with this information to prevent detrimental effects on plant growth.

Water quality is one of the least understood aspects in plant production impacting medium pH and nutrient availability. You need to understand your water quality and how it affects your growing medium.

Growing medium—it's with your plants from start to finish

The growing medium is the one factor that must support plant growth from the time of sowing or transplanting to final planting. When considering media, examine physical characteristics: porosity, aeration, water holding capacity and cation exchange capacity.

• **Medium porosity** determines the amount of oxygen available for root development, gas exchange and nutrient and water absorption. Aeration depends on container size. With shallower containers, the amount of water that can drain through the pot is less because of the lack of vacuum created by the force of gravity.

- **Air porosity** of media ranges from 0 to 2 percent in the plug cell to 15 to 20 percent in a 6-inch pot. Improve aeration by adding various particle sizes, such as calcined clay, sand-finished perlite and vermiculite.
- **Media water-holding capacity** affects watering frequency. Uniform distribution of a variety of particle sizes provides pockets for water retention and oxygen needed for root development.
- **The cation exchange capacity** of a medium (CEC) measures ability of the medium to prevent nutrients from leaching, thus retaining them for plant use. Peat-based mixes typically have a low CEC, usually requiring frequent fertilizer supplements. Chemical properties to monitor closely in a medium include pH and soluble salts.

> Your water test and media are fine, but the young leaves on your plugs are yellow and small.
> **What's the problem?**
> It is April—hot and dry with frequent watering. Calcium becomes available to the plant, boron is leached out of the growing medium and becomes deficient. This problem usually occurs on petunias and pansies. The solution for boron deficiency is one application of S.T.E.M. at half the recommended rate—as a corrective measure only.

Soluble salts in the starting medium should be .75 to 1.0 (based on a 1:2 soil: solution dilution ratio). Starting pH should be 5.8 for most crops and may raise to 6.5. Know nutrient concentrations in the medium, but even more important is knowing nutrient ratios in the medium and in subsequent tissue tests from the crop. Based on Iowa State University tests, the ratio should be 1:1:1:$\frac{1}{2}$ for nitrogen, potassium, calcium and magnesium, respectively. Take complete soil tests every two weeks and send them to a lab for analysis. If the ratios in the soil test are close, growers will usually produce a quality plant. Monitor pH and soluble salt levels weekly, on site. It is important not only to test the growing medium before using it, but also to trial it before incorporating it into production.

Observe the manganese to iron ratio, keeping it 1:2. If a mix has 50 parts per million manganese and 20 ppm iron, iron deficiency may occur for some plants. Understanding the ratio of these elements is more critical than knowing their specific concentrations.

Fertilizing for optimum growth

How do you decide which nutrients to apply? A liquid fertilizer program for most young plants includes nitrogen, potassium, calcium, phosphorus, magnesium and minor elements. Nutrients are usually obtained from ammonium nitrate, potassium nitrate, calcium nitrate, diammonium phosphate, urea and minor elements.

> One grower had a soil sample come back with results indicating that there should not be a problem. The calcium level was 150 ppm; the magnesium, 30 ppm. Both elements were in optimum range, but the plants were all yellow.
> **Why?**
> Because the calcium was high enough to interfere with magnesium uptake. The plant had magnesium deficiency: yellow middle and bottom leaves.

What type of fertilizer can be used? 20-20-20 has been a standard fertilizer for the greenhouse industry for many years, but many firms are switching to fertilizers such as 20-10-20 and others specifically formulated for soilless mixes. These formulations contain much higher micronutrient concentrations and less ammoniacal nitrogen. Humus forming from the decomposition of peat in soilless mixes ties up micronutrients. Thus, more micronutrients are generally applied to plants growing in soilless media.

Ammonium toxicity problems often occur in soilless media due to high levels of ammonium or urea. The 20-20-20 formulations invariably contain about 70 percent ammoniacal nitrogen, a rate too high for many plant species growing in soilless media. Formulations such as 20-10-20 Peat-Light Special contain 40 percent ammoniacal nitrogen, a safe level for year-round use. Growers have occasionally found benefits in switching from 20-20 20 to 15-15-15 fertilizer, even though it was applied at the same nitrogen rate. The benefit was likely due to the lower ammoniacal concentration—about 50 percent.

There's really little variation in the type of fertilizer necessary. Ammonium-based fertilizers—applied at concentrations varying from 25 to 350 ppm—yield soft plant growth. If harder growth is needed, apply 20-0-20, which contains calcium and potassium nitrate. Apply the ammonium form of nitrogen when soil temperatures are above 65 degrees F. Reports that ammonium causes toxicity on some plants usually occur in a cool, waterlogged soil, when the ammonium is converted to ammonia.

Changes in pH, soluble salts and nutrient concentrations in a plug medium from planting to maturity.*	Starting medium	Mature plug seedings
pH	5.8-6.5	6.2-6.5
Soluble salts	.75-1.2	1.0-1.5
Nitrogen	40-75	60-100
Phosphorus	10-15	10-15
Potassium	35-50	50-80
Calcium	50-75	80-120
Magnesium	25-35	40-60
Sulfate	75-200	75-200
Chloride	10-20	10-20

*Based on Iowa State University Tests.

Calcium and magnesium are frequently found in low concentrations in both soil and tissues. Why? The lower initial pH in a mix ties up calcium and magnesium. Calcium is needed for cell walls; magnesium for chlorophyll, which is involved in sugar production. If a grower observes a plant with small bottom leaves and light green in color, the cause may be magnesium and calcium deficiency.

If the proper nutrients aren't available to the plant in the first few days of growth, the plant won't achieve its potential growth. It's extremely critical to have optimum feed at proper ratios in the early stages of growth.

How to fertilize poinsettias

What are the fertilization requirements to produce an optimum poinsettia? The nutrient requirements for poinsettias are very different at different stages of development.

Highest feed is needed when bracts are just starting to form and initiate. Feed early, especially if you are direct sticking. Nutrient charge for the propagation medium of direct-stick cuttings should be low; 0.5 EC will promote faster rooting. Take care on a sunny day when soil starts to dry out, as harmful salts will accumulate around roots.

Fertilizing plugs

A grower can start feeding plugs at 25 to 75 ppm from the day of sowing. Use two fertilization programs during Stage 2: 20-10-20 or 20-0-20 Peat-Lite Special. For most plugs, maintain pH at 5.8 and soluble salts around 1.0. If the pH is below 5.5, up to half of the germination of some seedlings, such as cyclamen and impatiens, can be lost. In the later stages of plug growth, use 20-10-20 for optimum growth and 20-0-20 to tone the plant.

To develop a plug fertilization program, two major considerations must be taken into account. The pH needs to be between 5.8 and 6.0 for most elements to be available to the plant, and soluble salts should be approximately 1.0. Salts can be 1.5 at stages 2 and 3. During the final growth stage when the plug will be shipped, soluble salts should be 1.0. In a plug cell soluble salts go from approximately 1.0 to 1.5 to 1.0.

Poinsettia fertilization programs should include low ammonia concentrations until October 1, when phosphorus levels would be approximately 20 ppm. A 20-10-20 can be used, but discontinue this program after October 1 or when flower initiation starts.

After October 1, change to total nitrate feed such as 20-0-20. (Note that 20-10-20 does not contain calcium.) Some growers use only nitrate nitrogen—no ammonium—and it does work very well, although some ammonium with nitrate at early stages may be preferable. The amount of feed at October 1 and the next two to three weeks should be the highest EC level for that crop, falling between 1.5 and 1.8.

That is a critical time to supply maximum nutrients for the plant. Growers have observed larger bracts with optimum feeding programs during this stage of development, especially concentrating on the ratio concentrations of nitrogen, potassium, calcium and magnesium. Our studies have indicated that the ratio of iron to magnesium should be approximately 2:1. Based on work that has been done, drop 20-10-20 as early as mid-September, but certainly by October 1.

Today for most cultivars, some growers don't feed with ammonium. Why the emphasis on ammonium? Ammonium and sodium can interfere with calcium uptake and cause the cessation of terminal growth, resulting in a curled, cupped leaf or bract.

Even browning and necrosis on the leaf tip can be related to calcium deficiency, resulting from the effect of ammonium and sodium interfering with calcium uptake. Poor leaf expansion is usually caused by low calcium and low magnesium—all resulting from low starting pH in the media. The calcium problem can usually be corrected by adding foliar calcium fertilizer, such as calcium chloride, at a rate of 250 ppm.

For poinsettias, start with a low nutrient charge (.75 to 1.0). Monitor calcium and phosphorus to obtain the best root development possible. Drop ammonium from the feeding program by October 1. Maintain highest EC levels during initiation stage; EC should 1.0 after bract development. Reduce nutrient level during the last two to three weeks, but don't terminate your feed program. Make sure the EC is approximately 1.0 for the mature crop.

> A pot mum grower needs alkalinity of 100 ppm. His alkalinity is 350 ppm. What can he do?
> - **Problem**: Alkalinity in irrigation water: 350 ppm HCO_3. Wanted: 100 ppm. 250 ppm HCO_3 need to be neutralized in the water.
> - **Givens**: There are 4.04 meq/l of HCO_3, and 1 meq/l of HCO_3 = 61 ppm. 7 fl. oz. of 85 percent phosphoric acid neutralizes 1 meq/1.
> - **Solution**: Add 28.272 fl. oz. of H_3PO_4 to neutralize 4.09 meq/l of HCO_3.

"

David S. Koranski is professor of floriculture, Iowa State University, Ames.

Growing Ideas

March 1989

by P. Allen Hammer

Agribrom can provide useful benefits.

There is a lot of talk concerning the chemical Agribrom and what it does for plant growth. Greenhouse managers in Indiana continually ask me questions about the chemical. In my travels around the United States, I have seen several greenhouses using the chemical. They have all made very positive comments about Agribrom.

Several months ago we installed an Agribrom system in our mist propagation system at Purdue University. As with many growers, we had lots of questions about how and what to install. The unknown can be a little scary.

Agribrom is user friendly

Our experience with the system has been very positive. Fears of installation and proper working of the system have gone away because it is very simple and easy to maintain. We are applying Agribrom in the mist system with a small Dosatron injector set at 1:100. The concentrate solution cannot be made more concentrated than 1:100.

Before installation, I was concerned about the injector working on our mist system with our very low flow rate. We run as few as six mist nozzles for four seconds. The system has, however, worked without problems.

The small Dosatron is designed to function properly between a flow of 8 to 400 gallons per hour. The injector was installed in the supply line to our mist propagation area. We did install a bypass around the injector as a protection against a malfunction.

We are maintaining 5 to 10 parts per million (ppm) bromine in the water at the mist nozzle. An easy to use test kit makes it simple to test the level of bromine in the greenhouse. You can adjust the concentrate for the desired level, as well as check on the accuracy of the injection system. I wish we could check nutrient levels as easily as we can check the bromine level.

Fertilizer and acid require separate injectors

We make only enough concentrate for a day because the manufacturer recommends not allowing the concentrate to become "old." Agribrom should also not be mixed with fertilizer or acid in the **concentrate** tank.

If you are going to inject fertilizer or acid with Agribrom, a twin head or two separate injectors must be used. The concentrated fertilizer and/or acid will react with the Agribrom.

Although Agribrom is effective over a wide pH range, the concentrate for acid injection at 1:100 is generally below a pH of 2.0. Agribrom is not stable at that low of a pH. We are presently studying some ways to apply Agribrom and acid with a single injector.

The high pH (7.2) and hard water deposits of our mist system at the university certainly require acidification in addition to the Agribrom. Until we can determine a better system, two separate injectors should be used for the two chemicals to keep them from reacting.

Wiping out algae, slime and rot problems

What are the results from using Agribrom? The algae and slime problem in our mist propagation area has gone away. We have greatly reduced our cutting loss from "rots." We are seeing faster and better rooting on a wide diversity of plant materials. In a seed or cutting propagation area, I certainly think Agribrom can provide some very useful benefits.

We have not had our system installed long enough to study it for plugs, but with the problems we have experienced in the past it has great potential in reducing some of the plug problems. We will be giving it a real test this spring. An additional very positive result is that we have seen no phytotoxicity with any crops to Agribrom.

Start with a clean area

Like any system, the greenhouse manager can be a positive or negative influence on the success of the system. Agribrom will certainly not clean up an already messy propagation area, at least in reasonable time. It is important to start with a "clean" area.

If the chemical is not maintained on a constant basis at the correct concentration, you again cannot expect good results. I am convinced that in the proper hands it is a good tool for the propagation area. Our experience has been very positive.

"

P. Allen Hammer is professor of floriculture, Purdue University, West Lafayette, Indiana.

How to get 273 plugs out of a 273-cell tray

December 1988

by William C. Fonteno

Air and water content in a plug have four corners: medium, media handling, container and watering practices. By understanding these factors, you can think of them as management options and consider them as a package in trying to optimize air and water conditions in the plug.

E veryone seems to be searching for the ideal plug mix. The criteria are simple: A mix with good aeration (that doesn't dry out too quickly and can be used in all plug container sizes), contains all the nutrients necessary, and stores indefinitely. Sounds easy, right?

Such a mix is not impossible as long as price is no object. For about $100 per cubic foot, a medium can be designed that can provide all of these elements. All you need to do is make a medium that is not affected by other forces in the greenhouse.

I prefer to take a different approach. Think about the mix as a complement to these other forces in the greenhouse environment. Don't try to make the mix totally responsible for air, water and nutrition. This may sound a little strange at first, but if you look "inside" plug mixes, you will better understand what determines air, water and nutrition in a plug mix.

Test and compare commercial and homemade mixes

Before we start, let's take the issue of commercial versus homemade mixes. For my money, commercial mixes are preferable to mixing my own. At North Carolina State University, we have analyzed approximately 400 mixes, both commercial and homemade, over the last five years. There are good mixes and poor mixes on both sides. I have not, however, found a single home-grown mix that was really superior to the better commercial mixes available today.

This past summer I visited greenhouse growers in Denmark, Germany, Holland and Ireland. None of these growers were mixing their own media. Some have special blends, but this work is done by media specialists. Consistency, quality control and price make commercial mixes a better deal.

Since plug mixes are in their infancy, most plug producers are forced to make their own, but there are plug mixes that are commercially available. I urge you to try them. Test them and compare, but don't switch until you are satisfied with the performance.

Liquid feeding is best

The chemistry of nutrition in greenhouse crops can be complex. In plug production, nutrition is much more simple to handle. The general trend is to grow plugs with a lower nutrient component than other crops. A minimum base charge is placed in the mix and additional nutrition is applied as liquid feed. This gives the grower added flexibility in speeding up or holding plugs.

Liquid feeding also provides more uniform distribution of nutrients from cell to cell, tray to tray and house to house than incorporating nutrients into the growing medium.

A few words of caution about pH of homemade plug mixes: pH below 5.5 affects the germination of several bedding plant species. Although most mix recommendations call for a pH range of 5.8 to 6.2, it may take from 24 hours to seven days for the pH to adjust to this level. This depends on the ratio of mix components, particle size and grade of lime used, the salts used to make the base charge, and alkalinity of the irrigation water. We see more problems with low pH in grower-made mixes because it is generally used sooner after manufacture than commercial mixes.

Air and water in media are not fixed properties

Balancing the air and water content in media is one of the biggest problems facing plug growers. During early plug growth, cells are too wet and many seedlings drown. Later on, these cells dry out too quickly. These are inherent problems with plugs. You cannot eliminate them yet, but there are several things you can do to reduce the negative effects.

Growers tend to think of the mix as the overriding factor which determines air

**Figure 1
Factors affecting
air and water status
in plugs**

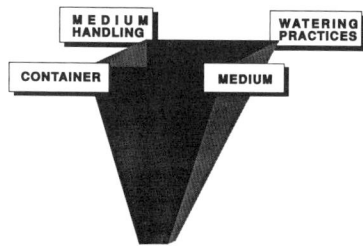

MEDIUM HANDLING WATERING PRACTICES CONTAINER MEDIUM

and water content in the root zone. Therefore, most mixes made over the last 20 years have been classed according to their air and water values as if they were fixed properties of the mix. They are not.

There are four major factors which affect the air and water status in plugs (Figure 1). I like to think of them as the four corners of a plug cell, each necessary in supporting the air and water content in that plug. They are:

- medium (components and ratios)
- media handling
- container (size and shape)
- watering practices

Small particle sizes will not improve air and water content

The medium influences the air and water content in the root zone. Most mixes used for plugs contain 30 to 60 percent peat moss. Generally, sphagnum is preferred because it has an advantageous fiber structure over hypnum or reed-sedge peats. This sphanum peat moss structure allows for good aeration and drainage. A few words of caution: simply using a sphagnum peat does not assure uniformity.

In Table 1 we can see the air and water content of several medium components in both 288 and 648 plug cells. The first three are all Canadian sphagnum peats (CS Peat I, II and III).

Although the Canadian peats provide similar air and water contents in large containers (which are not shown), they are different enough in water content and air space to perform differently in plugs that are on the same greenhouse bench.

Aggregates are generally added to peat moss to provide more rapid drainage and increase aeration. Most commonly, vermiculite, perlite and polystyrene beads are added. Vermiculite is the aggregate used most often and in the largest ratio, from 20 to 60 percent by volume.

The size of vermiculite commonly used in general potting mixes is Grade No. 2 (horticultural grade), which provides larger pores. Plug mixes, however, generally contain Grade No. 3 (which is finer) to allow the mix to flow more evenly into trays at filling. Ironically, Grade No. 3 is one of the poorest aggregates for adding air space because it holds less water and much less air. It is also more susceptible to compaction and structural collapse.

So which components contribute air space and which ones contribute to water content? Most plug mixes we have tested have between 2 and 4 percent air space in a 288 plug cell. Peat alone can give this level of aeration.

In Table 2, we combined peat and vermiculite—the two most common components—in three ratios in both 288 and 648 plug cells. With high ratios of vermiculite, we have our greatest drainage in both 288s and 648s, but we lost water holding capacity. Notice that the 1 peat: 1 vermiculite mix ratio was similar in both air and water content to the CS Peat I alone in Table 1. Increasing the ratio to 3 peat: 1 vermiculite did not really change the air content in either the 288 or 648 plug tray (Table 2).

What does all of this mean? First, peat plays a much more dominant role in plug trays than it does in larger plant containers. Second, aggregates may or may not help improve drainage depending on size and shape of particles. Third, smaller particle sizes don't necessarily improve air and water content. In fact, it can hurt.

Table 1
Which components affect air and water content?

Medium	288 Plug Container Capacity	288 Plug Air Space	648 Plug Container Capacity	648 Plug Air Space
CS Peat I	85%	2.6%	87%	0.7%
CS Peat II	84%	1.0%	84%	0.2%
CS Peat III	92%	1.9%	94%	0.4%
Vermiculite #2	64%	8.8%	69%	4.1%
Vermiculite #3	57%	0.1%	57%	0.0%
Perlite	53%	10.6%	59%	4.2%
Polystyrene	29%	6.2%	34%	1.0%

Table 2
Component ratio affects air and water content

Medium	288 Plug Container Capacity	288 Plug Air Space	648 Plug Container Capacity	648 Plug Air Space
1 Peat: 1 Vermiculite	85%	2.8%	87%	0.5%
1 Peat: 3 Vermiculite	74%	4.2%	77%	1.2%
3 Peat: 1 Vermiculite	87%	2.9%	89%	0.6%

Tall plugs are better than short plugs

The second corner of plug air and water content is container size and shape. The main reason why it can be harder to grow a good plug than a good pot mum is because of the plug cell itself. Plug cells have only two basic problems: they are too short and too small. Plug cells are so short that, at best, they drain very little, and, at worst (like in 648 waffles), they do not drain at all.

For example, the 1 peat: 1 vermiculite mix ratio shown in Table 2 has an air space of 2.8 percent in the 288 cell and 0.5 percent in the 648. This same mix

GrowerTalks on Plugs

has an air space value of 13 percent in a 4-inch pot and 20 percent in a 6-inch pot.

The importance of plug cell height is illustrated in Figure 2. A normal 273 plug is approximately 1 inch tall. We "manufactured" a tall 273 cell with the same length and width at the top opening, but made it 2 inches tall with the same general taper. The effect on drainage was dramatic.

The four mixes in Figure 2 are commercial plug mixes run through our lab for diagnostic purposes and are simply listed as Mix 1, 2, 3 and 4. Notice the difference in air content between the short and tall plugs. Air content went from a range of 1 to 3 percent in the short plugs to 5 to 10 percent in the tall plugs. If you could get 10 percent air space in all of your plugs, we could cut your plug production problems in half.

Figure 2
Short vs. tall
273 Plugs

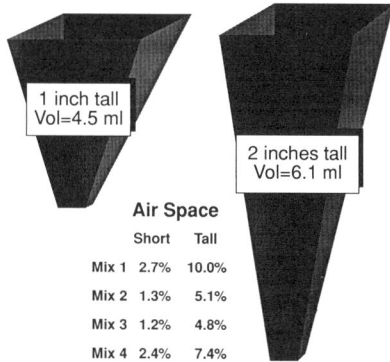

1 inch tall
Vol=4.5 ml

2 inches tall
Vol=6.1 ml

Air Space		
	Short	Tall
Mix 1	2.7%	10.0%
Mix 2	1.3%	5.1%
Mix 3	1.2%	4.8%
Mix 4	2.4%	7.4%

Ironically, plant production in small containers started out years ago in tall cells. One of the factors which contributed to the success of the Speedling company in Florida was the Todd flat. Vegetable transplants were grown in relatively large cells that were about 3 inches deep. Even when smaller cells were later developed, they were still quite tall.

I'm not sure we truly appreciate the effect of cell height because today most plugs are grown in short cells. Obviously, it is possible to grow good quality plugs in these short plug cells. But the smaller the cells, the greater the chance of the plants being overwatered or underwatered.

The only way to be economically competitive in the future is to decrease our costs and increase our efficiency. It is not efficient to spot water trays individually because some dry out faster than others. Nor is it efficient to guarantee 200 plugs in a tray that holds 273 cells.

Part of this inefficiency is due to conditions changing too fast in plug production. We need to increase plug cell height and media volume to reduce this rapid fluctuation and allow for more automation.

Square cells are better than round cells

Another issue is container shape. Good plugs can be grown in both round or square plug cells. Round cells are actually portions of a cone, while square cells are sections of pyramids (Figure 3). Of the two designs, the square cells are preferred because they have a larger volume.

In Figure 3, we see a round and a square 288 cell. If you calculate the volumes of each you find that the round cell has a volume of 4.66 milliliters, while the square cell has a volume of 6.18 milliliters. Although these numbers are small, the square cell is 33 percent larger than the round cell.

Figure 3
Round vs. square

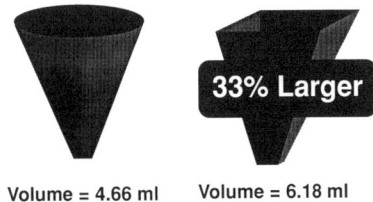

33% Larger

Volume = 4.66 ml   Volume = 6.18 ml

288 Plug Cells

The extra volume translates to more water being available to the plant and less chance of drying out. This extra volume does not necessarily increase air space percentage. As long as the height remains the same, however, there is no decrease in drainage, so air space is not adversely affected.

Most plug tray designs today are square instead of round. While round cells are still on the market, there are no advantages to growing in round cells. Even if you can get a better price on round trays, the decrease in media volume is not worth it.

Don't pack medium or stack trays while handling

The handling of a mix can greatly affect the air and water content of that mix. Frankly, many mix manufacturers were, at first, reluctant to produce a plug mix because of the problems of grower handling and liability. Their efforts to select the best material, conscientiously blend them, carefully package them, and ship them can be undone on the other end by someone who handles the mix improperly.

One factor in handling is compaction. Plug trays should be lightly filled and the excess brushed away. This can be done by hand or machine flat fillers. The medium should not be packed down and trays should not be stacked directly on top of each other. You can cut your air space in half and even eliminate air space by compaction.

Add water to the mix before filling cells

The second consideration is moisture content of the mix prior to filling trays. When water is added to dry components, such as peat, they hydrate and swell. This swelling helps to create more aeration by reducing the tendency of the particles to nest within one another. This effect is not so dramatic on larger containers, but can be the difference between success and failure of a plug crop.

Most plug mixes tend to be inadequately moistened prior to tray filling. In fact, most growers would prefer to put the mix in dry and add water later. Unfortunately, this is like closing the barn door after the cows have gotten out. Water should be added to the mix before it is placed into the cells.

How much water do you add to the mix? For peat-based media used in pot plant production, approximately 100 percent by weight. Plug mixes should have approximately 200 percent by weight. This level will seem much wetter than normal, but will actually improve aeration. A few words of caution: do not soak down the mix after seeding. Light misting for germinating the seed is fine, but the mix does not need more moisture after filling.

Prefilling plug trays and letting them dry out can be detrimental if the trays are handled much before they are remoistened. Continuous flow mixers can also cause problems when portions of the mix sit on the conveyor belts overnight. Slight separation of the mix can occur on the belt which can result in a different mix in several flats.

Keep media mixing time to a minimum

If you are mixing your own media, the type of mixing action and mixing time is important. Batch mixers, which use paddles or hammers, tend to break down the structure of the peat. Augers, ribbon blenders or rotary mixers do less damage. In any case, mixing time should be minimum.

In a batch mixer, I prefer to mix components dry for one minute, add the base charge, mix again for one minute, add 200 percent moisture (by weight) through a high pressure mist nozzle as I mix for three more minutes.

The person at the end of the hose controls your profits

How you water a plug mix influences air and water content in the root zone more than the mix itself. Because plugs do not drain well they are easily overwatered. There is an old saying in the greenhouse business: "The person at the end of the hose controls your profits." This is certainly true for plugs.

Knowing when to water is perhaps the most important skill for a plug grower. It is also the biggest headache.

There are two basic options to watering plugs: "over the top" and from the bottom. Top watering can be by hand, boom or mist line. Subirrigation can be accomplished by mats or ebb and flood. Watering from the bottom generally provides more aeration. This is because water moves by capillary action from the bottom and never clogs pores used for air space.

Top watering actually floods the pores on the way through the container reducing aeration and causing overwatering. Much more care is needed to top water properly. The disadvantage with subirrigation on a mat is that the roots tend to grow into the mat. Subirrigation on ebb and flood benches eliminates that problem but can cause another.

Because the plugs are so short, the water level necessary to use the flood bench virtually saturates the plug cells. Therefore, if your plugs are about 1-inch tall, subirrigation can give you the same results as top watering.

You can produce very good plugs using subirrigation; it is not yet superior, however, to top watering given the height of plug cells in today's commercially available plug trays.

Grow on the dry side, not the wet side

Many growers produce plugs on the dry side. This practice produces a harder plug and reduces the chance of overwatering damage. This has as much of an effect on air and water content as the mix itself.

To illustrate this, I was asked to look into the media problems associated with a new plug tray design by a national supplier of plugs. The problem was that the plugs stayed too wet.

After some research, I contacted the company with my recommendations. I found out that the plug producer had already solved the problem by switching from a "wet grower" to a "dry grower."

The company had been growing small plants for years with few problems—until they switched to a smaller plug cell design. Their first approach to the problem was to redesign the medium. They reduced the problem, however, to a manageable level by changing growers (watering practices).

By understanding these four corners—medium (components and ratios), media handling, container (size and shape) and watering practices—you can think of them as management options. They should be considered as a package in trying to optimize the air and water conditions in plugs.

"

William C. Fonteno is associate professor of floriculture, North Carolina State University, Raleigh.

Know your media
the air, water and container connection

March 1988

by William C. Fonteno

We've changed the way we grow. Now it's time to change the way we look at artificial media and the containers we grow in.

The shift to smaller pot sizes for mass markets and the increased demand and use of plugs, as well as concerns about contaminated runoff water leaving greenhouses, encourages us to take another look at media, containers, watering practices and media handling.

Have you ever wondered why the same mix you use for poinsettias just doesn't work as well for bedding plants or in plug trays? In the visual presentation following, the factors which determine media air and water content are examined.

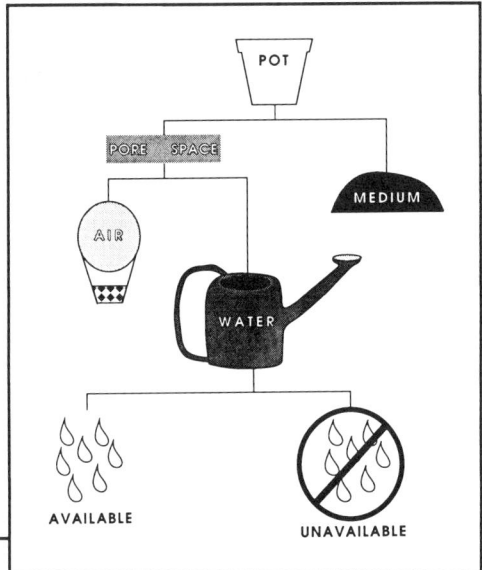

The root environment is made up of three main factors: the container, the solid portion of the medium and the pore space the medium creates. A medium's pore space holds either air or water. Mineral soils are about 50 percent solids and 50 percent pore space. Organic media have less solids but more pore space. Today's good commercial mixes have about 65 percent to 90 percent pore space. Water held in the medium may be either available or unavailable to the plant. Unavailable water is so tightly held to medium surfaces it cannot be removed by roots. In mineral soils, very little of the water is unavailable to roots. The amount of unavailable water in organic media or in commercial mixes is high.

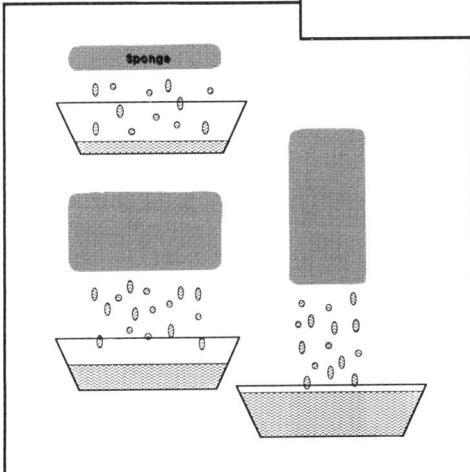

Very little water drains from a water-soaked sponge lifted horizontally. Tilting the sponge on its side increases drainage, but holding the sponge vertically—creating the tallest possible column—causes water to drain the fastest. This same principle is true with containers. The taller the container, the better the medium drains. Shorter containers not only drain slower, but they also contain less air.

The smaller the container, the less air and water it holds. Observe that in a 6-inch pot, chart 2 left, most of the commercial mixes listed have at least 20 percent air space. When these same media are put in a 273 plug tray, air content decreases to 2 percent to 4 percent. As the air con-

Container size affects air and water content

Container		Metro 360	Fafard 4	Ball II	Jiffy Mix	ProGro 300
6"	Water content	60%	64%	60%	73%	62%
	air	23%	17%	24%	20%	22%
4"	Water content	67%	70%	67%	79%	68%
	air	16%	11%	17%	14%	16%
Bedding plant cell (48/tray)	Water content	73%	75%	74%	85%	74%
	air	9%	6%	10%	9%	10%
273 plug tray	Water content	80%	80%	81%	90%	80%
	air	3%	2%	4%	3%	4%

tent decreases in smaller containers, note how water content increases, to at least 80 percent in most cases. The smaller the container, the more critical watering practices become.

Plug mix and container combinations

Container		Peat/ vermiculite	Coarse plug mix	Medium plug mix	Fine plug mix
Bedding plant cell (48/tray)	Water content	80%	64%	71%	79%
	Air content	7%	10%	4%	2%
273-cell plug tray (1" tall)	Water content	85%	70%	75%	81%
	Air content	2%	4%	1%	0.3%
273-cell plug tray (2" tall)	Water content	82%	65%	72%	80%
	Air content	6%	9%	3%	1%
648-cell plug tray	Water content	87%	72%	75%	81%
	Air content	0.3%	1.8%	0.2%	0.1%

Plugs either too wet or too dry? They are too wet because cells are too shallow to drain properly and they are too dry because cells are such small volume. In the comparisons charted below, note that the coarser the mix the greater the drainage and air content. In the 273-cell plug tray example, when cells are 2 inches deep, air content doubles or triples. Also documented in the illustration, tiny waffle cells do not drain well—attaining only 1.8 percent air content with the coarsest plug mix.

Filling/packing affects air space*

	6" standard pot	4" standard pot	Bedding plant cell (48/tray)
Light packing			
Available water	43%	51%	58%
Unavailable water	21%	21%	21%
Air space	23%	15%	9%
Medium packing			
Available water	44%	52%	56%
Unavailable water	26%	26%	26%
Air space	15%	9%	4%
Heavy packing			
Available water	45%	49%	52%
Unavailable water	30%	30%	30%
Air space	9%	4%	2%

*Medium used in tests: Peat/vermiculite.

Air and water content results from a combination of three factors: (1) pore space and available water as defined by the properties of the media; (2) the size and shape of the container; and (3) how the medium is handled prior to and during filling the container. Packing media into the container reduces air space and creates more unavailable water to the plant. The smaller the container, the more critical aeration becomes and the lighter it should be filled.

Here's why you shouldn't fill and stack plug trays or bedding plant flats

Container	Coarse plug mix	Medium plug mix	Fine plug mix
Bedding plant cell (48/tray)			
Filled and brushed			
Available water	45%	48%	42%
Unavailable water	19%	23%	27%
Air space	10%	4%	2%
Filled, pressed and refilled			
Available water	37%	40%	41%
Unavailable water	27%	29%	32%
Air space	3%	0.3%	0.2%
273-cell plug tray (1" deep)			
Filled and brushed			
Available water	51%	51%	54%
Unavailable water	19%	23%	27%
Air space	4%	1%	0.3%
Filled, pressed and refilled			
Available water	40%	41%	42%
Unavailable water	27%	29%	32%
Air space	1%	0.0%	0.0%
273-cell plug tray (2" deep)			
Filled and brushed			
Available water	46%	49%	53%
Unavailable water	19%	23%	27%
Air space	8%	3%	1.5%
Filled, pressed and refilled			
Available water	38%	40%	42%
Unavailable water	27%	29%	32%
Air space	3%	0.1%	0.1%
648-cell plug tray (waffle trays)			
Filled and brushed			
Available water	53%	52%	55%
Unavailable water	19%	23%	27%
Air space	1.8%	0.3%	0.1%
Filled, pressed and refilled			
Available water	40%	41%	42%
Unavailable water	27%	29%	32%
Air space	0.3%	0.0%	0.0%

Growers don't fill 6-inch pots with mix and then press other pots on top of them. Why is it such a common practice for plug and bedding plant trays? Note in the comparisons above that the lighter the media is packed, the more air space. In plug mixes, the coarser the mix, the more air space and the least amount of unavailable water. A fine plug mix heavily packed eliminates any chances of drainage and aeration.

"

William C. Fonteno is associate professor, North Carolina State University, Raleigh.

Feed plugs early

January 1988

As reported to Debbie Hamrick by David S. Koranski

Nutrition is important to young, growing plugs and as Dr. David Koranski has found through research at Iowa State University in Ames, with begonias and petunias the earlier you start, the faster the bedding crop grows.

Q: Early stages—are you talking about germination?

A: I am talking about the time you sow the seed, which we call Stage I. This is the stage that the seed is starting to germinate—the root emerges, but the cotyledons haven't emerged yet. Stage I is the time that growers should fertilize many of their plugs.

Q: What day is that?

A: That's day one or day two. In fact there can even be times when you fertilize the day you sow.

Q: And at what ppm?

A: That's going to depend on the nutrient charge of the medium, the cation exchange capacity of the medium, light intensity, temperature and certainly the cultivar you are growing.

If you have a high nutrient charge in the growing medium, less fertilizer is needed from day two. From our experiments, with a moderate nutrient charge in the medium, growers should feed at 50 ppm.

Q: What formulation are you recommending?

A: For the first stage, which normally happens from day one to day five, we suggest potassium nitrate or 20-10-20 peat-lite special. The reason for using potassium nitrate is that it may overcome dormancy and there is dormancy in petunias and other seed.

You have to be careful with 20-10-20 in not providing too much ammonium.

Q: Before, you didn't recommend starting to fertilize at this early stage, you said to start later.

A: Before we had not really made any recommendations. The reason for it was that we were using growth chambers at the time. It was difficult to fertilize in growth chambers, so feeding would start when the plugs came out and started to develop true leaves. That was anywhere from day eight to day 10 or 12, that was our suggestion. I was concerned about making a recommendation on fertilizing in the early stages because of potential salt buildup at rates that we would normally feed a plug in Stage II or Stage III (150 ppm to 200 ppm). I've always felt that those rates were too high to feed early stages.

Q: Should an early fertilization program be used with a standard plug media?

A: The program can be used with a standard plug mix. A soil test should be run on the mix. If the nutrient charge is higher than 200, the initial feeding program may be less than 50 ppm.

The pH of the medium for any plug program should be 5.8 and be maintained between 5.8 and 6.5.

Q: What about water testing?

A: Water testing should be conducted to determine pH and bicarbonate levels. Plug growers should also look at the concentration of boron (should be less than 5 ppm), chloride (should be less than 30 ppm) and the sodium absorption ratio (should be less than 2 ppm).

Q: You said that the initial fertilizer concentration depends on the nutrient charge of the media, crop, temperature, light intensity, etc. Can you relate any of these to common plug crops?

A: The most responsive crop to early fertilization has been petunias. The dry weight of petunias may increase as much as 50 percent with early feeding. It's dramatic.

Q: It works really well for petunias. What are some other plug crops this program will work well with?

A: Well, almost equally well and maybe even better for fibrous begonias.

One of the difficulties with begonias is rate of growth. Our experiments have shown that fibrous begonias have an accelerated growth rate with this program. One of the difficulties with begonias is maintaining uniform growth during the first three weeks. Many times, begonias stall out at the end of this time—they won't grow anymore. If you take a soil test, in many cases you will find that the soil is very low in nutrients.

The plants are using the fertilizer and as much as 50 percent to 75 percent of the nutrient charge will be leached with the first watering. Plants need to be fed, if not on a continuous basis, then one to two times a week. The early fertilization program usually prevents plants from stalling out.

Q: How many weeks is it going to cut crop time by starting to fertilize early?

A: For petunias it's approximately a week to a week and a half. For begonias, it may be a least two weeks.

Q: Will it cut crop time off geraniums?

A: The program will probably not reduce crop time, but plants should be larger and more compact.

Q: By fertilizing petunias and begonias early you can definitely cut weeks off crop time?

A: This has been proven in our lab and with many growers throughout the country. The feeding program may overcome dormancy and provide more uniformity in seed germination and seedlings.

By applying a fertilizer containing calcium, growers should also get a more vigorous root system.

Q: Is early fertilization continuous?

A: The early fertilization program is still visual observation. If the plant is very dark green or starting to stretch, we may not feed it for a couple of days. A fertilizer program as well as a growth regulator program, as far as concentrations and frequency are concerned, is something that is a grower's art as well as a skill.

You have to look at and observe the plant. In conjunction with observation you are doing a soluble salts and pH reading at least every week or two. That way you know your pH is in the optimum range and your soluble salts are also optimum.

Q: You will use this early, low concentration program until when? Day 8? Day 10?

A: The program is implemented until about day 10. Then growers usually increase the fertilizer program up to the normal rate at this time, which is somewhere between 150 ppm and 350 ppm depending on the crop.

"

David S. Koranski is professor of floriculture, Iowa State University, Ames, and Debbie Hamrick is editor of GrowerTalks *magazine.*

Growing ideas

December 1987

by P. Allen Hammer

"I am sure many greenhouse fertility problems can easily be corrected by proper use and calibration of the fertilizer injector."

It is interesting how often I get calls—after receiving soil samples—asking if I think the fertilizer injector is causing a fertility problem. Although I firmly believe in soil sampling, it is not the way to test your injector for accuracy. Calibration of the fertilizer injector should be a routine job in the greenhouse. The importance of an accurate fertilizer injector in greenhouse production cannot be overstated. It is the one piece of equipment that is in constant use. And your crop productivity and quality are dependent on the accuracy of your fertilizer injector.

If you wait until you "see" plant growth problems from a malfunctioning injector the damage to productivity and plant quality can be very costly. There are many methods used to calibrate the injector; however, we find the following method to be easiest and probably most accurate, assuming you have correctly mixed your fertilizer concentrate.

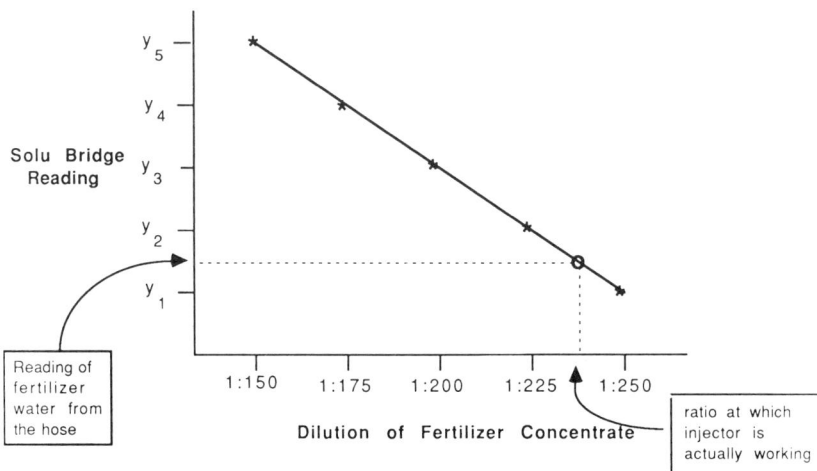

Figure 1. SoluBridge reading at various fertilizer dilutions.

Useful conversions

1 fluid ounce = 2 tablespoons = 29.6 milliliters or cubic centimeters
1 cup = 8 fluid ounces = ½ pint = 16 tablespoons = 236.5 milliliters or cubic centimeters
1 gallon = 4 quarts = 16 cups = 128 fluid ounces = 3.785 liters
1 pound = 16 ounces = 453.6 grams

The fertilizer injector works as a dilution device. A 1:100 ratio means that one part of concentrated fertilizer solution is mixed with 99 parts of water to give 100 parts of dilute fertilizer solution at the hose. Or, if the ratio is 1:16, then one part of concentrated fertilizer solution is mixed with 15 parts of water to give 16 parts of dilute fertilizer solution at the hose. Therefore, for calibration, we make several solutions of known ratios in the range of our injector. For example; if we are calibrating a 1:200 injector, we make solutions of 1:150, 1:175, 1:200, 1:225 and 1:250. These solutions need to be made very carefully. It does not matter what units of measurement you use, just that you use the correct ratio. I suggest either milliliters or fluid ounces. Measuring equipment can be purchased from many scientific supply houses or may be available at a local drugstore or hardware store.

Be sure to use enough volume to avoid as much mixing error as possible. For example, a 1:150 solution could be made with 10 milliliters of concentrated fertilizer solution mixed with 140 milliliters greenhouse tap water for a total volume of 150 milliliters. Note that greenhouse tap water is used to make solutions. This will account for naturally occurring salts in your tap water.

Once all the appropriate solutions are made, a SoluBridge is used to measure the soluble salts in each solution. The readings are then plotted on a chart as in Figure 1. Note that the plot should be a straight line. If it varies a great deal, do the dilutions again and use more care in your mixing. Make a separate chart for every fertilizer you use in the greenhouse.

Once this graph is obtained it is easy to determine if your injector is working properly. Simply take a water sample at the hose and measure the soluble salts—the reading will tell you if the dilution is correct when compared to Figure 1. You can also tell what ratio the injector is working at if the reading is different than expected and adjust your concentrate accordingly.

One real advantage for the system is you can take samples from different houses and hoses when using one injector for multiple areas. This also makes it easy to check the injector accuracy at different flow rates. In using such a system, take water samples while you are watering. This gives you an idea of what the plant is really getting from the fertilizer system.

I would also raise the question of when was the last time you calibrated your SoluBridge? Just because you get a reading does not mean it's correct. A standard reference solution can be made by dissolving 0.744 grams of dry potassium chloride (certified pure) in distilled water and dilute to 1 liter. This solution should read 1.41 millimhos/cm on the SoluBridge. If your instrument reads differently, read the manual for possible solutions.

I cannot overemphasize the importance of checking the accuracy of your fertilizer injector. It is important that checks be made regularly of the solution being applied to the plant. An injector may work very differently when a hose is "wide open" as compared to watering a single bench with a spaghetti system. This method makes such checks very easy. I am sure many greenhouse fertility problems can easily be corrected by proper use and calibration of the fertilizer injector. Check your injector today.

99

P. Allen Hammer is professor of floriculture, Purdue University, West Lafayette, Indiana.

Growth Regulators

Research Update

February 1990

by John Erwin and Royal Heins

"Don't use B-Nine on your vegetable plugs—it's illegal. Try temperature for height control."

Vegetable transplant production represented 8 percent of the bedding and garden plant market in 1988. One of the major problems in producing vegetables as a bedding plant is height control. Traditionally, growers control vegetable height using growth retardants and water and/or nutrient stress. B-Nine was the only labelled growth retardant for use on vegetable transplants, but the controversy surrounding B-Nine use on food crops has resulted in withdrawal of the registration of B-Nine on any food crop. What can you do to control vegetable transplant height without B-Nine?

You can try temperature to control height of some vegetable plants. Temperature has a dramatic effect on plant stem elongation. Specifically, as the DIFerence (DIF) between day and night temperature (day temperature minus night temperature) increases, stem elongation increases.

DIF works on vegetables

We grew vegetables under three temperature environments to provide a range of DIF values. The temperature environments were 73 F day temperature/63 F night temperature (positive DIF), 68 F day and night temperature (0 DIF) and 63 F day temperature/73 F night temperature (negative DIF). Each of the different environments had a 12-hour light period, so average daily temperatures within each environment were the same.

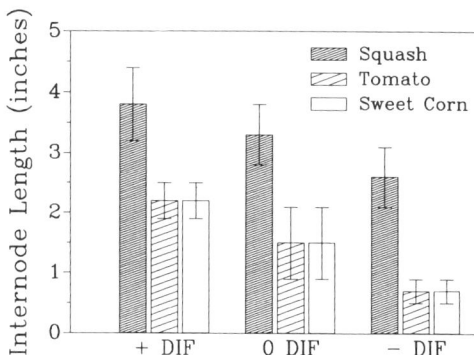

Vegetable crops listed in Table 1 respond to DIF. Pea and cucumber were the only vegetables treated that didn't respond, although different pea and cucumber cultivars grown last year responded to DIF. This suggests that some pea and cucumber cultivars may respond to DIF and some may not. We have heard similar reports with petunias also.

Constant temperature a problem?
Try early morning cool pulses

The data presented are from plants that had constant temperatures during the day and night. Constant temperature in the greenhouse, especially during the day, is difficult. Temperatures typically rise in the afternoon. If this is the case in your greenhouse, you may want to consider a cold temperature pulse during the first two hours of the morning.

Research on Easter lily and annual flowering plants suggests that plants are the most sensitive to temperature during the first two hours of the day. If you are able to drop early morning temperatures below the night temperature during the first two to fours hours of the day, you'll reduce stem elongation dramatically.

For maximum reduction in stem elongation, drop temperatures during the morning hours as close to first light of day as possible. As daylight increases, a cool temperature pulse treatment has less effect on reducing plant height. Maintaining cool temperatures as long as possible during the day further increases plant response.

One final word: It is illegal to apply any growth retardant to vegetable transplants. Beyond the liability associated with an illegal application, the potential damage to the entire bedding plant industry is immense. Don't use B-Nine on your vegetable transplants this spring!

Table 1: Vegetable plug response to day and night temperatures.

Plant	DIF	Internode length (inches)
Sunny tomato	+	2.2
	0	1.5
	-	0.7
Watermelon	+	1.0
	0	1.4
	-	0.2
Squash	+	3.8
	0	3.3
	-	2.6
Sweet corn	+	2.2
	0	1.5
	-	0.7
Bean	+	1.9
	0	1.6
	-	1.4

"

John Erwin is assistant professor at the University of Minnesota, St. Paul. Royal Heins is professor at Michigan State University, East Lansing.

Keeping pansies short in the plug flat

July 1989

by Shawn Laffe and Roger Styer

Meeting fall demand for pansies means growing plugs in July and August, when temperatures are high and stretching ruins plug flats. Growth retardants—applied correctly—can control plug height.

Pansy plugs for fall sales must be grown during the hot summer months of July, August and September. Warm temperatures (75 degrees and higher) cause pansies to stretch quickly, promote soft growth and blast flowers. Even the most extensive cooling systems for greenhouses have difficulty maintaining 75 degrees F or lower in August. Chemical growth retardants are often necessary to supplement environmental controls.

The response to chemicals of different species and varieties varies. In general, the more vigorous the variety, the higher the chemical rate or frequency of application is necessary. The large flowered Majestic Giant pansies are more vigorous than the smaller flowered Crystal Bowl series. Colors within a series respond differently, too. Blue or yellow pansies are generally more vigorous than orange or red pansies.

Fine tuning pansy plug chemical needs

The summer of 1988 was especially long, hot and dry. Most pansy plug producers had to resort to chemical growth regulators. More than a few hard-luck stories surfaced: Plugs either defied chemical control or were permanently dwarfed by a chemical overdose.

Because of these problems, we conducted a study at Ball Seed Co. comparing the effects of chemical growth retardant on pansy plugs. We examined effects of Bonzi, Sumagic and B-Nine on two different pansy cultivars (Crystal Bowl Orange and Universal Orange) and stages of pansy plugs grown under controlled environmental and cultural conditions.

We single-seeded pansies into 406-cell plug trays containing Sunshine No. 3 mix, watered in and lightly covered with coarse No. 2 vermiculite. Germination took place at 72-degree F soil temperature in a germination chamber having fine mist and fluorescent lights. We moved germinated seedlings to another controlled environment chamber for growing on, with 72-degree F soil temperature and continuous fluorescent lighting (600 footcandles).

Plugs were fertilized as needed with 50 ppm 20-10-20 Peat-Lite Special and 25 ppm calcium nitrate. We made foliar application of Sumagic, Bonzi and B-Nine at the first, second, third or fourth true leaf stages, or at multiple stages. All chemicals were applied at the volume rate of two quarts per 100 square feet.

We measured plug height (petiole length) at the time of transplant (four weeks after sow date). Acceptable treatments (based on plug height and appearance) were transplanted into flats and finished in a greenhouse at 55- to 65-degree F soil temperature. We observed plants for residual effects of growth regulator treatment on size and time to flower.

Treatments producing the most acceptable plugs when grown at 72 degrees F in our chamber.	
Crystal Bowl Orange	**Universal Orange**
Sumagic 1 ppm, leaf 2	Sumagic 1 ppm, leaf 2
Sumagic 3 ppm, leaf 3	Sumagic 3 ppm, leaf 3
Bonzi 1 ppm, leaf 2	Bonzi 1 ppm, leaf 3
Bonzi 3 ppm, leaf 3	
B-Nine 2500 ppm, leaves 1+2+3	B-Nine 2500 ppm, leaves 1+2+3

In some cases, higher chemical concentrations or multiple applications aren't necessary. In those situations, we opt for the lesser treatment to conserve the amount of chemical used and time needed for re-applications.

All chemicals give good control

Results show that all chemicals affect pansy plug height compared to untreated plugs, and that both cultivars respond similarly. For Universal Orange, Sumagic at 3 ppm produces more control than 1ppm for all leaf stages (Figure 1). The amount of control decreases as leaf stage increases. Multiple applications made at the first and third or second and fourth leaf stages produce greatest plug height control.

Plugs treated with 1 or 3 ppm Bonzi are shorter than plugs treated with Sumagic (Figures 1 and 2). Generally, Bonzi at 3 ppm produces shorter plugs than 1 ppm. The effectiveness of Bonzi on plug height decreases as stage of development increases. Multiple applications produce the greatest control at both Bonzi concentrations.

As with Sumagic and Bonzi, the controlling effects of B-Nine applied in multiple applications (first, second and third true leaf stages) also increase as the chemical concentration increases (Figure 3). B-Nine doesn't cause plug stunting, although 5,000 ppm will produce highly compact plugs.

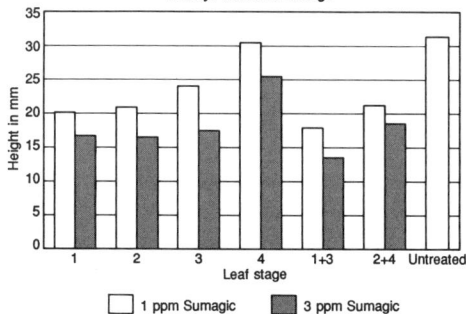

Figure1
Effect of *Sumagic* on pansy plug height when applied at different leaf stages
Variety: Universal Orange

Timing is everything

In pansy, plug height control is mainly dependent on time of chemical application. Plugs treated at the first true leaf stage with Sumagic or Bonzi at 1 or 3 ppm will be stunted.

Multiple applications of either compound also stunt plug growth. Chemical applications at the fourth true leaf stage don't control pansy plug growth.

We determined desirable plug growth visually as approximately 3/4-inch to 1 inch in height with a compact stem, expanded leaves and a well-developed root ball.

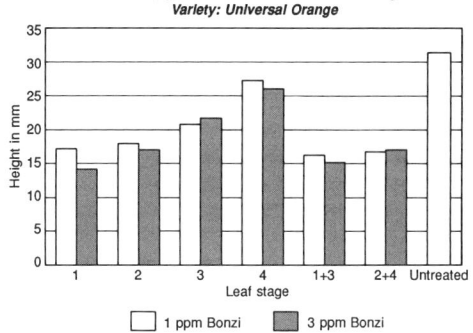

Figure 2
Effect of *Bonzi* on pansy plug height when applied at different leaf stages
Variety: Universal Orange

Besides plug height, all chemicals affect leaf color and growth habit. Sumagic and Bonzi restrict leaf expansion compared to untreated plugs, but enhance green color and root development, making a "tighter" plug. B-Nine enhances leaf expansion and foliage color, producing a more "open" plug, but doesn't increase root development.

We transplanted plugs from acceptable treatments into cell packs and finished them in a greenhouse. No difference was observed in time to flower, flower size or plant habit at flowering compared to untreated plugs.

Environment regulates responses

Growing conditions play an important role in the effect of chemical growth regulators. Warm temperatures promote vigorous growth and make plants less responsive to chemical controls. Conversely, plants growing under cool temperatures may not require as much chemical treatment, if any at all.

Under the conditions of this study, certain applications of Bonzi and Sumagic controlled pansy plug height acceptably and didn't delay flowering or cause stunting. It is possible, however, that these chemicals would have had a greater effect at lower temperatures.

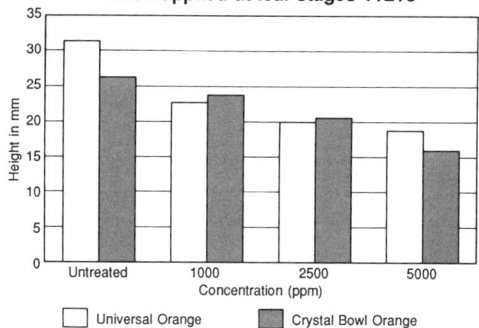

Figure 3
Effect of *B-Nine* on pansy plug height when applied at leaf stages 1+2+3

B-Nine also produces desirable control of pansy plug growth. At higher growth temperatures, B-Nine loses effectiveness as the rate of plant growth increases and may not be able to control plant growth.

Trial chemicals to work out rates

Fertilizer, moisture and light intensity also affect plant growth and its response to chemicals. Because each and every greenhouse environment is unique, growers must trial any chemical on a **small scale** and keep records of calibrations, environmental and cultural conditions and plant response.

Finally, growers need to determine what chemical, concentration, time and frequency work best in their operations. Sumagic and Bonzi are volume-critical chemicals. If more solution is applied, more control will result.

Both chemicals are effective on plant stems and in the root medium. Excessive run-off from the stems or foliage can create a greater controlling effect.

These chemicals are safe to use! Properly treated pansy plugs grow out without any delay! Bonzi and Sumagic just take more effort on the part of the applicator in monitoring the environment and application techniques.

"

Shawn Laffe and Dr. Roger Styer are in Plant Research and Development at Ball Seed Co., West Chicago, Illinois.

Control impatiens height in the plug stage with Sumagic or Bonzi

February 1989

by Roger C. Styer

With powerful chemicals like Bonzi and Sumagic, growers can now control height on previously uncontrollable crops—like impatiens. There's increasing concern, though, about the effects of these chemicals on later growth and questions about application procedures.

Growers **can** get an early start on impatiens height control with Bonzi and Sumagic, providing they know how to use these difficult-to-apply chemicals. By treating impatiens plugs, growing impatiens dry to control height may become a thing of the past.

Medium to high concentrations at four weeks give best control

Five parts per million (ppm) Sumagic or 20 ppm Bonzi control plug height, with greatest control occurring with a double application at two and four weeks after sowing. This double treatment, however, tends to be too stunted, and plugs do not grow out very well.

For effective height control without reducing growing out capabilities, Bonzi at 20 or 30 ppm and Sumagic at 5 or 10 ppm, applied two weeks after sowing, works best. At this stage, seedlings are just starting first true leaves.

Both chemicals are active in the root zone as a drench; residual action in the medium may be responsible for height control when applied two weeks after sowing. With later applications, such as four weeks after sowing, Sumagic controls plug height better than Bonzi. Under the experimental conditions, plugs start stretching rapidly after four weeks, and neither growth retardant controls height when applied five weeks after sowing.

Chemicals promote roots and branching

Regardless of chemical concentration or time of application, both Bonzi and Sumagic increase impatiens root growth. Greatest root response with Sumagic occurs with an application four weeks after sowing.

This particular treatment's effect on roots, combined with excellent height control for a four-weeks' application, demonstrates Sumagic's effectiveness when used later in plug development. The benefit to growers who transplant this four-week-treated plug is a well-developed, short plug with lots of roots that take off faster.

BONZI
Plugs - Week 2
Flat - No Treatment

10 ppm   20 ppm   30 ppm   Control

12-22-87

SUMAGIC
Plugs - Week 2
Flat - No Treatment

2.5 ppm   5.0 ppm   10.0 ppm   Control

12-22-87

Both of these growth retardants also increase basal branching in the plug stage. This branching effect continues after transplanting, resulting in a well-branched finished plant.

Treated plugs grow out as superior plants

Morphologically, impatiens growth is altered when Bonzi and especially Sumagic are applied two weeks after sowing. Treated plugs have a deep, green color, more roots and thickened leaves and stems that give a toned appearance without the use of stress. Upon transplanting, treated plugs continue to grow, producing a well-developed, basally branched finished plant that flowers on time. High concentrations, repeated applications or improper application rates (greater than 2 quarts per 100 square feet), however, cause flowering delays. Different conditions may also give different results. Each grower must experiment individually with these compounds to determine the best way to apply them. **99**

Dr. Roger C. Styer is in Plant Research and Development at Ball Seed Co., West Chicago, Illinois. The author acknowledges Laurette Lech for her technical assistance on this project.

Plug height control: Choosing the chemical that's right for you

February 1989

by Jim Barrett

When it comes to plug height control, B-Nine is the safest chemical to use, but is inactive on key crops, like impatiens, geraniums and pansies. Where, when and exactly how should you use A-Rest, Cycocel, Bonzi and Sumagic?

The production of a high quality plug is one of the most challenging tasks in the greenhouse industry. Successful plug growing requires a high degree of control over all factors in production that affect seed germination and seedling growth.

Controlling seedling size is one of the big challenges, because when growth is at an optimum stem elongation can be excessive. Many plug producers use growth retarding chemicals for obtaining a rapidly developing seedling that is short and stocky.

Using growth retardants successfully on plugs is more difficult than it is on finished bedding plants. There is a greater chance of overtreatment that can result in stunted seedlings that are very slow to start growing or do not resume growth after transplanting.

To be successful, make sure that the best chemical is used, proper rates are applied and correct spray procedures are followed.

Are you growing "hard" or "soft" plugs?

There are generally two types of plugs: "hard plugs" that have been grown slowly and "soft plugs" that have been given optimum conditions for growth. Hard plugs are commonly produced by growing or finishing at low temperatures. Drought stress and low fertilization also contribute to hard plugs, along with heavy use of some growth retardants.

Generally, hard plugs are tougher and withstand shipping better than soft plugs. Also, hard plugs can be held in the plug tray longer before they have to be transplanted. They are usually slower to start growing after transplanting, however, and finishing time will be greater than with soft plugs. Whether a plug is grown soft or hard also affects the growth retardant regime required to produce the plug.

Chemical rate depends on growing conditions

The proper rate for a given chemical depends on several factors. Growing temperature is probably the most important. When temperatures in stages 3 and 4 are kept at or below 60 degrees F, seedling growth is slow, stem elongation is reduced, and growth retardants generally are not needed. If 65-to 70-degree F temperatures are used, a growth retardant is often needed to reduce stem elongation.

Likewise, stem elongation in most species is reduced when plugs experience drought stress. Where this is practiced, the amount of chemical required is reduced. Some species, such as impatiens, are very responsive to fertilization and will stretch if the level of fertilizer is too high, which leads to a high chemical requirement.

Optimum chemical rates for plugs are lower than are used on the finished crop—often one-fourth to one-half that used on cell packs or 4-inch pots. The best strategy is to make repeated applications of a low rate. This reduces the chance of over treatment, and treatments can be discontinued if a growth retardant is not needed.

Time of application varies with stem elongation. In general, the first treatment is made when the stem starts to elongate in early stage 3. Then subsequent applications are made as needed at five- to seven-day intervals.

Which growth regulators work best with plugs?

B-Nine is the most commonly used chemical for finishing bedding plants and is widely used on plugs, as it's safe and there's little chance of overstunting the plugs. Rates vary from 750 to 2,500 parts per million (ppm).

The problem with B-Nine is that it is not very active and many plug growers find that it does not provide adequate control on some of the more important species such as impatiens, geraniums and pansies.

Cycocel is used on seed geraniums because it causes early flower initiation. It is active on only a few species and causes leaf chlorosis or yellowing on all species, which is not acceptable to some growers.

A-Rest is a good chemical that is active on a wide range of species. It is not used on finished bedding plants because it is very expensive to use as a spray compared to the other chemicals. Plug growers may be able to afford A-Rest in special cases because of the high relative value of the crop.

A starting point for A-Rest is 5 to 25 ppm. At this rate, A-Rest costs 20 cents to $1 per tray, compared to less than 5 cents for other chemicals. A-Rest fills a position between the less active chemicals, B-Nine and Cycocel, and the more active and difficult to use chemicals, Bonzi and Sumagic. Thus, growers should evaluate A-Rest on crops not controlled by B-Nine.

Bonzi and Sumagic are the two new growth retardants that are highly active on all species and have the highest inherent risk. When used correctly they provide the best height control, but if a mistake is made, plants may be overly stunted.

Application techniques are critical for these chemicals. The spray must be applied uniformly over the crop and from one spray time to the next in order to obtain consistently uniform results. The effect of these chemicals is closely tied to the amount of spray that is applied, so that when higher volumes of spray are used, the effect from a given rate is increased.

Using optimum rates for Bonzi and Sumagic is critical. The optimum rates vary widely between growers because of environment, growing procedures and application techniques. They also vary for different species and different cultivars within a species.

The optimum rate for Bonzi is between 5 and 20 ppm. The most sensitive crops are geranium, vinca and pansies, which would require the lowest rate. Rates of 8 to 10 ppm are a good starting point for most other species.

Sumagic, introduced in 1990, is very similar to Bonzi except that it is slightly more active. Sumagic rates are between 3 and 20 ppm, with the sensitive crops needing lower rates. For trials, rates between 5 and 10 ppm should be used on most plug crops.

For crops that are not so responsive to Bonzi, such as salvia and snapdragons, Sumagic may be a better chemical. With vinca, Bonzi can cause spotting on leaves, but Sumagic does not. Fibrous begonias are very sensitive to Bonzi **and** Sumagic; the above rates may cause severe overstunting. Bonzi and Sumagic should be used on begonias with extreme caution.

Undecided? It's a question of cost and control

A number of growers have used Bonzi with mixed results. There have been several cases of plugs being treated with Bonzi and then being shipped to another grower who could not get them to grow. News of these kinds of failures has contributed to a reluctance by some growers to try Bonzi. There are, however, several plug growers who are very successful with Bonzi and are able to take advantage of its high activity on crops not controlled by B-Nine.

Bonzi and Sumagic can be used by plug producers to hold down stem elongation of pansies, impatiens and other difficult crops. Growers should do their own trials to determine correct rates for their growing situations and to ensure that their application techniques give uniform results. Bonzi and Sumagic should not be used on a large scale until the growers are confident of responses that will be obtained.

B-Nine is the safest growth retardant to use on plugs and is probably the chemical of choice for those species it affects. For other species, growers are left to make a choice between the expensive chemical, A-Rest, and a chemical that is more difficult to use, Bonzi or Sumagic.

"

Jim Barrett is Professor of Horticulture, University of Florida, Gainesville.

Day/night temperature combinations can hold plug height

December 1988

by Vic Ball

The cool day/warm night plot is thickening! *GrowerTalks* has reported that the major help in height control on lilies and poinsettias is cooler day than night temperatures. Now new data, again from Michigan State, reveals more surprising and very helpful responses of both plugs and bedding plants to varying night and day temperature regimens.

Regulating night/day temperatures can control plug height

Adjusting night/day temperature will cause striking differences in height (stretch) of plugs! If the day is warmer than the night, plugs of almost any species grow taller. As the day temperature is cooled relative to the night temperature, plugs grow shorter. The cooler the day relative to the night, the shorter the plug.

For example, 60 degrees F day, 65 degrees F night (researchers say "negative temperature difference" or "negative DIF") means short, compact plugs. Sixty-five degrees F day, 60 degrees F night (positive temperature difference or positive DIF) means taller plugs.

THE EFFECT OF DAY AND NIGHT
TEMPERATURE ON
POT PLANT DEVELOPMENT

DAY 71
9 HOUR DAY
IMPATIENS WALLERANA

25/15   15/25

Examining the effects of day and night temperatures on *Impatiens wallerana* development, researchers grew plants under different temperature regimes. Impatiens grown with 77 degrees F days and 59 degrees F nights were larger than plants grown with a day/night temperature combination of 59/77 degrees F. Cool days and warm nights favor compact growth.

Constant days and nights avoid chlorosis

With most plants, cool days/warm nights cause chlorosis of the young, expanding leaves. The chlorosis normally clears up as the leaves mature under negative DIF conditions or within a couple days when plants are exposed to a positive DIF. The major concern with chlorosis on plugs is that plug growth can be slowed due to reduced photosynthesis when young plants are chlorotic.

Therefore, John Erwin, researcher, recommends that plugs not be exposed to large negative temperature differences early in seedling development. Instead, he suggests a constant (similar) day and night temperature for plugs most of the way through production. "A constant temperature difference results in short, compact plugs with no chlorosis," he says.

Understanding day/night temperature responses are important to growers, whether they grow plugs or almost any of our crops.

Bedding plant responses to temperature

Impatiens grown from transplant to sale with warmer day versus night temperatures will stretch. With cooler day versus night temperatures plants are compact and bushy, which is the type of plant most growers want.

The difference is striking! Responses are much the same for petunias and other annuals. A slight delay in flowering occurs with petunias grown under cool days.

Unfortunately, so many bedding plants are grown in marginally ventilated poly houses that tend to heat up excessively on sunny spring days and often are heated only sparingly at night due to fuel costs. The resulting temperature combination is warm days with cooler night temperatures, which means rapid stretch. We now understand why petunias stretch so rapidly in April!

It must be pointed out that poly allows maximum heat from the sun (infrared) to enter during the day, heating plants. At night, untreated poly allows the maximum escape of heat from plants (except for IR films, which trap infrared rays).

In fact, some growers may well opt for their less fuel costly, cool nights plus B-Nine applications to control plant height. It is an option.

Cooling at sunrise can "cool" the whole day

Deep South growers with those warm early spring days are hard pressed to provide cooler days. Research at Michigan State offers a suggestion.

Tests so far show that if temperature is kept cool the first one to two hours after sunrise, then even though the mid-day temperature may rise, the net effect will be at least a cooler day. Balanced with as warm a night as practical, you get much of the cool day/warm night effect. Pulling thermal sheets open abruptly at sunrise is another way to get a cool first hour or two.

In most cases, John counsels to run as near a constant day and night temperature as possible for most bedding plants. The cooler the constant temperature, the slower the growth and flowering, but also the better the quality. Of course, impatiens and begonias need warmer temperatures than crops like petunias for normal growth.

"

Vic Ball is editor in chief of GrowerTalks *magazine.*

Calibrating sprayers for Bonzi

April 1988

by Don C. Wilkerson

Several new plant growth regulators have recently been developed for commercial greenhouse crops. Bonzi is the first to reach the market and now has an expanded label for use on bedding plants. It is extremely effective in controlling the height of a wide range of plants, however some problems have been reported concerning the predictability of response.

Bonzi is so active that growers literally have the opportunity to shut down plant growth for extended periods of time. This can be a very effective tool for controlling plant height, but in some situations has resulted in delayed or unsaleable crops. Although much of this variability has been associated with the rate of chemical used, little emphasis has been placed on the "volume" of material applied.

In general, the term rate refers to the amount of chemical required to make a given concentration of material (i.e. $1/2$ ounce of Bonzi/1 gallon of water=63.5 ppm). Volume, on the other hand, is used to describe the amount of that material applied to a given area (i.e. the Bonzi label recommends 2 quarts/100 square feet). The relationship between rate and volume is extremely well defined in the area of herbicides. However, since these chemicals are used sparingly in the production of greenhouse crops, many producers are unfamiliar with potential problems that may occur.

Much of the variability reported from the use of Bonzi is directly related to the amount of active ingredient applied per plant. For example, a $1/2$ ounce/1 gallon rate of Bonzi applied at 2 quarts/100 square feet will deliver 0.60 milligrams of active ingredient/square foot. This same rate applied at 1 quart/100 square feet will deliver only 0.30 milligrams of active ingredient per square foot. Based on this information it should be apparent that both rate and volume must be carefully controlled to obtain predictable results.

Careful calibration and precise application techniques can greatly improve the effectiveness of these new "super" plant growth regulators. However, applying chemicals in most greenhouse situations presents some unique problems that field applicators do not have to contend with. Realistically it may be difficult, if not impossible, to consistently apply the exact amount of chemical desired.

Calibrating spray equipment

Although there are several methods used for calibrating spray equipment, the following procedure outlines a basic approach:

1. Start sprayer, adjust pressure, and collect 2 quarts of liquid from the spray gun. Record the time required. Repeat this operation for increased accuracy (T = Time).

2. Multiply the time required (T) to deliver 2 quarts of liquid by the bench area to be treated (BA = bench area to treat). Now divide by 100.

$$\text{Spray Time Required} = \frac{\text{TxBA}}{100}$$

Example: From the calibration procedure it takes nine minutes to collect 2 quarts of liquid from the spray gun. The bench area to be treated is 6' by 30' (180 square feet). Now plug these figures into the formula.

$$\text{Spray Time Required} = \frac{9 \times 180}{100}$$

$$= 16.20 \text{ minutes}$$

Based on these calculations, if the pressure is maintained the same as during the calibration procedure, the area to be treated will require 16.20 minutes of spray time to deliver the equivalent of 2 quarts/100 square feet of bench area. The key to uniform plant growth regulator application is to maintain a steady spraying motion and speed over the entire area to be treated. With repeated applications this procedure should become routine.

Plant growth regulators such as Bonzi are rapidly becoming necessary tools for the production of quality greenhouse crops. If the potential of these powerful new chemicals is to be maximized, growers must continue to refine their application techniques. By reducing the margin of error producers will realize greater predictability and value for their products.

"

Don Wilkerson is extension floriculturist at Texas A&M University, College Station.

A tour of growth regulators on geraniums

March 1988

by E. Jay Holcomb

New information on geranium response to growth regulators is becoming available, but it is also important to review the information currently available. Since this review is an overview of research results from many researchers' publications, it is not intended to provide growers with specific recommendations.

Gibberellic acid

In general, gibberellic acid promotes cell elongation,, and increases the number, size and rate of opening of flowers. Although gibberellic acid does not have a major beneficial effect on geraniums (*Pelargonium X hortorum*), there are effects of gibberellic acid on Regal geraniums (*Pelargonium X domesticum*). A series of studies was conducted at Penn State to determine if gibberellic acid could influence the flowering of the regals. Gibberellic acid was sprayed at 0, 5, 50 or 500 ppm weekly for 4 weeks. Gibberellic acid applications did not increase the height of the Regals, but reduced flower bud abortion and increased the number of inflorescences and the percentage of plants that flowered. Apparently, gibberellic acid applications may be substituting for the cold temperatures that usually encourage flowering on Regal geraniums. In summary, gibberellic acid can cause stem elongation and delay flowering in geranium, however, on regal geraniums, gibberellic acid applications do not cause stem elongation and actually promote flowering.

Growth retardants

Growth retardants reduce or restrict internode elongation and at very high concentrations may reduce leaf size. Commonly, applications of growth retardants will result in darker green leaves. Most growth retardants delay flowering, however some will promote flowering. There is evidence that growth retardants provide some resistance to certain environmental stresses.

● *Cycocel*

Under high light conditions, Cycocel applications have been shown to increase inflorescence number on geraniums by increasing the number of shoots per plant, and to reduce time to flowering. With low light conditions, these effects of Cycocel diminish. Since Cycocel, like many growth retardants, act as anti-gibberellin compounds in the plant, the promotion of flowering may be the result of Cycocel affecting gibberellins in the plant.

Cycocel can be applied to geraniums as a drench or as a spray. The rate and timing of Cycocel applications are important. Foliar sprays of Cycocel at 1,000 ppm beginning 14 days after sowing, produce short and early-flowering plugs. Even low rates (250 ppm) have shown to be effective in producing earlier flowering and shorter plants in hybrid geraniums. With Cycocel, the time of application is very important and very early applications (two to three weeks after sowing) are effective in controlling height and inducing early flowering.

One factor of concern to the consumer is the garden performance of Cycocel-treated plants. Generally, Cycocel-treated plants are shorter in the greenhouse, and subsequently remain shorter in the landscape.

Cycocel applications will also reduce height on many cultivars of ivy geranium (*Pelargonium peltatum*). In general, a foliar spray of Cycocel at 1,500 ppm will reduce elongation, and multiple applications—up to three applications—will provide additional growth reduction. The Balcon family of ivy geraniums does not seem to be responsive to Cycocel.

● *A-Rest*

A-Rest has not been shown to be very effective in controlling the height of geraniums or ivy geraniums.

● *Bonzi*

A new growth retardant that shows promise for use on geraniums is Bonzi. Bonzi applied as a spray at 10 to 30 ppm restricts hybrid geranium height compared to untreated plants. As a drench, a concentration of 1 to 3 ppm is sufficient to cause significant height control on geraniums.

There is concern from growers about the length of time geraniums can be held in cell-packs without stretching. We treated cuttings in cell-packs in late August with Bonzi at a rate of 250 or 2,500 ppm one time or 25 or 250 ppm three times with application one month apart. In December it was obvious that the higher rates had virtually stopped stem elongation. The cuttings were repotted into 4" pots and grown until March. By March, the plants treated with the higher rates of Bonzi still had not stretched. Bonzi is a very active growth retardant and its effect will persist for a long period of time.

● *Sumagic*

Our studies indicate that Sumagic, like Bonzi, is very active and very persistent. For geraniums in pots, a foliar spray of 25 ppm provided a great deal of height control compared to nontreated plants. In packs, a 5 to 10 ppm spray was all that was necessary. Because Sumagic and Bonzi are very active, growers must use these materials with care and precision.

The aforementioned results are from research programs and the materials and methods used may not be labelled uses of the materials; thus growers should not use this information as recommendations for growth retardant applications. Mention of a product does not constitute endorsement of that product. Always read and carefully follow label directions.

"

E. Jay Holcomb is associate professor of floriculture, Pennsylvania State University, University Park.

Controlling plug height with Sumagic

January 1988

As reported to Debbie Hamrick by David S. Koranski

The new growth regulator Sumagic from Chevron is active on a wide range of bedding plant species and cultivars.

At Iowa State University, David Koranski and his team have conducted tests on Sumagic and its effects on controlling bedding plant and plug height. Following, David outlines some of the points he has learned from working with the new chemical.

The basics

The chemical structure of Sumagic is almost identical to the chemical structure of Bonzi. Both Sumagic and Bonzi are gibberellic acid inhibitors. Gibberellic acid is the plant growth hormone that stimulates stem elongation. Both chemicals also cause leaves of treated plants to become darker green—taking on a green sheen.

But even with similarities, there are major differences. Based on David's research, primarily conducted with Sumagic, he has found Sumagic to be more consistently uniform when applied to bedding plants. "What I mean by that is, you put it on at a certain rate and volume and it seems to have the same response time after time."

Sumagic seems to show response on a wider range of bedding plant crops. And, David's work has shown that it increases plant root growth.

When and how much?

Like many other growth regulators applied to bedding plants, Sumagic shows the best activity at the 2- to 3-centimeter leaf diameter stage for plugs and anywhere from the 2- to 5-centimeter leaf diameter stage for plugs transplanted into cell packs. (One inch is equal to 2.54 centimeters.)

As for rates, David says they should be worked out at individual locations depending on temperature, light intensities and bedding plant cultivar.

David has found that in petunias, doubles require more of the chemical than blues, blues more chemical than whites and whites more chemical than reds. So, within petunias, even the color of the cultivar makes a difference.

While rates vary, based on his research, David has found Sumagic to be active on many different bedding plants. "I've yet to find one bedding plant crop that is not responsive to Sumagic."

Petunias are very responsive. "It has been reported that Sumagic is very effective in controlling the height of petunias in the plug tray as well as in cell packs. You're looking at rates as low as 5 parts per million and as high as 20 parts per million."

David suggests that it's best to start out at low concentrations and to apply more frequently. "Start out at maybe half the recommended rate or less than suggested and apply more frequently." And remember, plug trays take a much less concentrated spray than plants in the flat.

"What we have to be very cautious about is that compounds available today—Florel, Cycocel, or B-Nine—can be used at 700 parts per million to 5,000 parts per million." With Sumagic, effective concentrations are in a much smaller range. "I don't really find that a problem. As professional growers today, we should be able to apply the correct volume of chemical to a given area."

Will plants grow out of Sumagic?

With gibberellic acid inhibiting growth regulators, some growers have found that their effects are long lasting. If you're spraying petunias in the plug tray, will they grow out of it when you transplant?

"That's a valid question and I think that problem can come if you put on too high a concentration. You can decrease the flower size by as much as 20 percent to 50 percent, the time to flower will be increased and at higher concentrations, some plants may not outgrow the growth regulator treatment."

How should growers avoid flower delay and decreasing flower size? "Conduct a test with plugs sprayed at different stages of development. Plant them out and see what happens."

Some researchers have shown that gibberellic acid sprays have overcome the dwarfing effects of Sumagic, but David doesn't see that as a viable solution to too high concentrations for commercial growers. "We don't have the time to work the crop with two, three or four different growth regulator applications. That's the benefit of a new growth regulator for bedding plants—Sumagic is a broad spectrum growth regulator, effective on many different crops," Dave says.

Trial is better than error

Try Sumagic on a limited scale to get a feel for what it will do for you. When new tools are used properly, it is possible to take some of the guesswork and luck out of dealing with Mother Nature.

"

David S. Koranski is professor of floriculture, Iowa State University, Ames, and Debbie Hamrick is editor of GrowerTalks *magazine.*

Equipment

Automation on a shoestring: Building a plug popper

May 1989

by Paul King

Mechanical plug poppers are not always cost efficient for small growers who transplant a limited number of plugs. Here's how—from hardware to precision drafting—one greenhouse owner built a plug popper for under $50.

There is no question that in the past 10 years the continuing increase in the use of plugs has been the most significant cultural change in the bedding plant industry. The huge plug operations are proof of their validity. Much has been written about growing plugs, but very little has been said about removing the billions of plugs used from their trays.

Machines now exist to remove and transplant plugs mechanically, with more under development. But for most smaller growers it is a chore that often becomes an expensive, time-consuming nuisance. Commercial plug poppers are neither very complicated nor inordinately expensive. But for our limited needs we believed that we could build an LHAH—loving hands at home— device to do the job at a fraction of the cost of a commercial model.

This past winter we set out to build a reliable, inexpensive popper. We studied what had to be done, and following is how we built our popper.

Plywood forms the popper base

We drilled two sheets of $3/4$-inch plywood, using a sheet of plywood as a back-up board to carry the pins that would dislodge the plugs. I suggest the top and bottom sheets be $3/4$-inch plywood for good vertical alignment—the back-up board can be just about anything. We used a $1/4$-inch piece on our 288 popper and $3/4$-inch on the 273, and both worked fine.

A few tests showed us the pins had to have a little play so they could dislodge the plugs by entering the drainage holes, which are not always in the center of the cell. We settled on $5/16$-inch and $1/4$-inch pins. Springs were to be used to hold the sheets just far enough apart so the tops of the pins would be just above the surface of the top board.

When pressure was exerted on the top board and an attached tray, the pins would enter the cells and pop the plugs. When the pressure was released the pins would recede. A few tests showed us the principle was sound and it became a matter of "do it."

As so often happens with a do-it-yourself project, particularly the first one, unforeseen problems arise. This project was no exception. How to locate the centers of the holes to be drilled became a real problem. A tray was useless as a template as the drainage holes tend to wander. We realized then we needed some precision drafting.

Using computer-assisted drafting

I have two daughters, both of whom are architects, and to me drafting is a person at a drafting table with tee square and rulers inking lines to produce a precise drawing. Fortunately I met Matt Coleman who heads drafting instruction at Buchholz High School in Gainesville, Florida. He showed me I was back in the days of "Sunny boy," and this was a perfect problem for CAD—Computer Assisted Design.

The pieces needed to make the plug popper work are simple and, once constructed and gathered, easy to assemble. This drawing shows a schematic view of the popper's working parts.

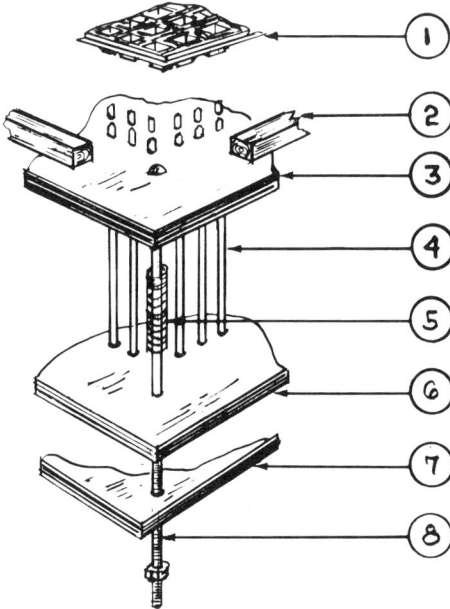

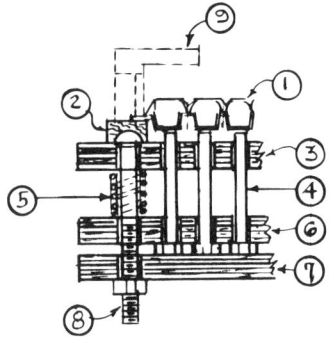

Corner detail

1 Plug tray

2 Lip support and positioning guide

3 Top sheet $3/4$" plywood—drilled

4 Dislodging pins— hex head bolts

5 Separating spring

6 Bottom sheet—$3/4$" plywood—drilled

7 Back up board— plywood of choice

8 Guide—$4^1/2$"x$^1/4$" carriage bolt and nut

9 Pressure frame

In 10 minutes, more or less, he had a printout of 288 cross hair drilling points more precise than anything produced by conventional drafting. It didn't take any longer to make a 273 configuration, which we also needed. This is not a difficult drafting problem, and any competent draftsperson should do it in an hour or less.

We locked the two pieces of $3/4$-inch plywood approximately 16 inches by 26 inches (this figure should not be smaller but can be longer) and taped the printout template to the top sheet. With a sharp-pointed scribe we carefully made a mark for the drill bit, using a $5/16$-inch Brad Point drill bit to prevent walking, and we drilled the 288 holes.

It took two of us about $1^1/2$ hours to carefully drill the holes. The more accurate the drilling, the better the action. Under no circumstances use a hand-held drill. You just can't hold vertical alignment.

Our next problem was finding a pin that would work in both poorly made and well made trays. A threaded carriage bolt, $2^3/4$-inches by $1/4$-inch worked fine in a well made precision tray, but tended to hang up in poorly made trays. Dowels worked fine, but anchoring them became another project.

A little searching found a 3½-inch by ¼-inch hex head bolt with approximately a ⅞-inch thread and a 2⅝-inch smooth shank. We put five or six at a time in a vise and sort of sliced the threads off with a hack saw. It was not a particularly hard problem; we did the cutting in less than an hour.

Putting it all together

Assembling the popper is simple. We bought four springs for about 50 cents each at a hardware store. We used four, 4½-inch by ¼-inch carriage bolts and nuts as spring holders and guide pins. We also used them to level the boards and tension the springs.

It is necessary to have a thin strip of wood placed under the tray lips to carry a pressure frame. Also, these strips can be used as position guides.

The accompanying drawing shows the basic pieces needed to make it work. The pressure frame and lever arrangement are the bells and whistles, which can be as simple or elaborate as you choose to make them.

We did find it became necessary—when we ran into bad trays—to have a piece of wood hold down the center of the tray. It just takes an instant to pull the plugs where the wood comes down. We also set up a lever arrangement. This is independent of the popper, and occasionally we move the device and use hand pressure to activate the plug popper.

For just under $50 for a template, plywood, pins and springs for material—plus a few hours of time—we have a popper that has been foolproof.

It takes our transplanters about 10 to 12 seconds to place a tray, pull the lever and remove the tray. Removing plugs is no longer a time consuming chore or nuisance. If you have any questions, feel free to call (603) 267-6554.

”

Paul King is the owner of Adhara Nursery, Gilmanton, New Hampshire.

The do-it-yourself seeder

April 1985

by Paul King

This is the recounting of one small business's search and its discovery of how to get into plugs without a sizable fixed investment or giving up a substantial piece of the pie. The solution wasn't perfect, but it worked well for us at Adhara—a small, profitable, wholesale vegetable bedding operation in central New Hampshire.

Three years ago, when it had become fairly obvious that plugs had a decided cost advantage in a wholesale operation, we realized we would have to automate our seeding or we would just wither away.

There is a place for an automatic seeder if you have the volume. In certain operations, particularly if you realistically allocate costs, it is cheaper to buy than grow. We don't have the volume (we have 5,000 feet under plastic) for the first option, and the second choice was not economically acceptable if we were to remain competitive and maintain our profit.

Working out the bugs

Early on (Adhara was established six years ago), we developed a vibrating seeder resembling the common battery model. It was powered by an electric razor attached to sort of an open funnel with a baffle. It worked well, but it was not suitable to seed plug trays.

The search began with a careful analysis of just what an automatic seeder does. An automatic selects a seed from as mass and then drops it at a precise location on a seed tray. How it is done and at what speed may vary, but that is what is accomplished.

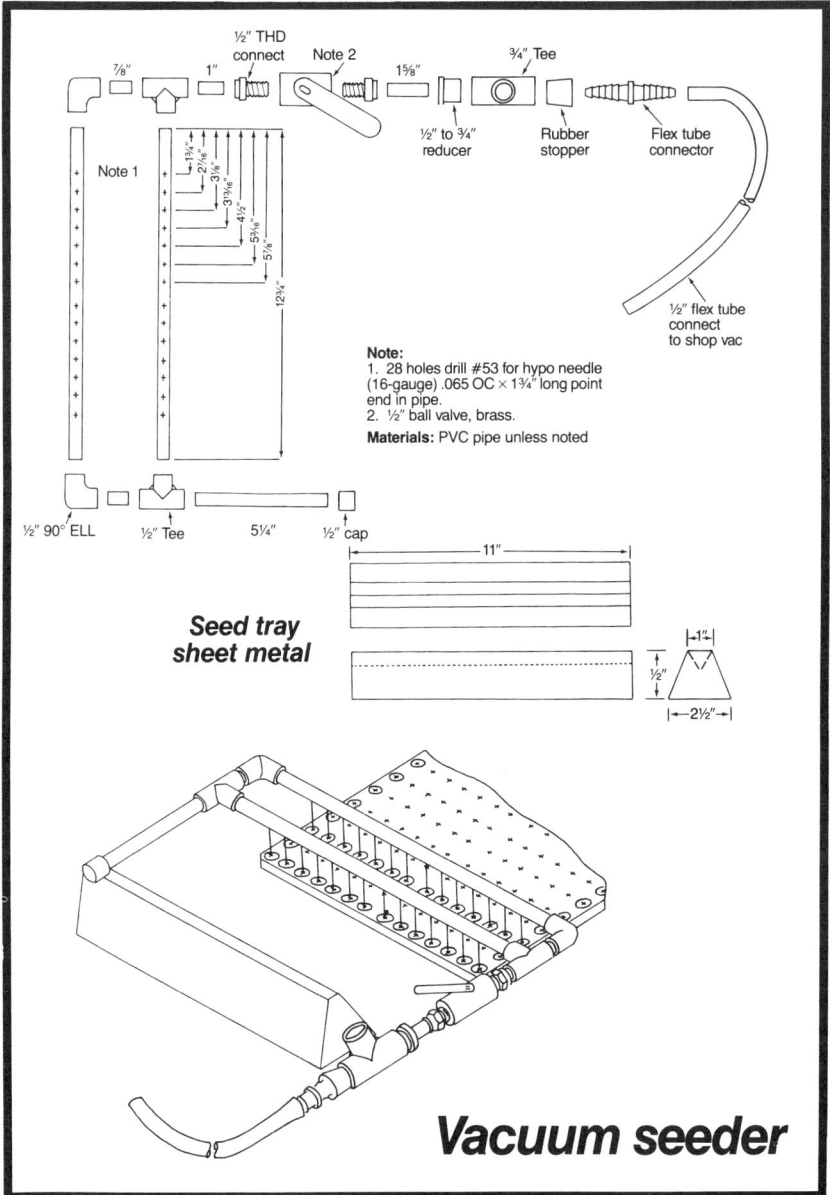

½" THD connect
Note 2
¾" Tee
1"
7/8"
1⅝"

½" to ¾" reducer
Rubber stopper
Flex tube connector

Note 1

1¾"
2⅛"
3/16"
3¹³/16"
4½"
5⅜"
5⅜"
12¾"

½" flex tube connect to shop vac

Note:
1. 28 holes drill #53 for hypo needle (16-gauge) .065 OC × 1¾" long point end in pipe.
2. ½" ball valve, brass.
Materials: PVC pipe unless noted

½" 90° ELL
½" Tee
5¼"
½" cap

11"

Seed tray sheet metal

1"
½"
2½"

Vacuum seeder

We set out to find a way to duplicate this function quickly, yet inexpensively. Many leads (mostly dead ends) were pursued. Luck also played a role. A 25 cent plumbing tee was an important breakthrough.

After a number of tries, we had a crude, simple, and inexpensive device that worked. Also, it was assembled from common "off-the-shelf" plumbing and medical items. The first seeder was a row of hypodermic needles (points down) inserted in a piece of hard PVC pipe in a closed system and attached to a vacuum cleaner.

Once the principle was proven, refinements came along. The two-handled "U" shape made the seeder easy to use. A valve to regulate the vacuum pressure for different seed was an important improvement.

We now have a semi-automatic seeder with no moving parts that is a horticultural tool—not a Rube Goldberg gadget.

The finished product

The seeder handles the seed of tomatoes, eggplants, peppers, and all the cold crops very well. Tests with detailed marigold seed have gone well, and we plan to grow more this coming season.

We can seed a 20-row seed tray with a 45-needle single bar or a 406-plug tray with two 14-needle bars in less than two minutes. Since our production is mainly short and intermittent runs, we average about 15 trays per hour.

Our germinating chamber space is limited, and we run approximately 60 percent seedlings and 40 percent plugs. This rate costs out well, and a few hours a day of seedling let us match seedling maturity to our transplanting schedule. We most likely will build another germinating chamber soon to increase the plug ratio.

The principle is simple. A vacuum is established at the tip of each needle. The needles are brought in contact with a mass of seed and each needle picks up and holds a seed. Then, still under vacuum, the seeder is moved with the seed held at the needle tips. When the needles are at the proper position, the vacuum is broken and the seed drop into the tray.

A problem was to "make and break" is done by rolling the fleshy part of the hand at the base of the thumb off and on the open tee in the right handle. It can just as well be in the left handle, if desired.

Just about anyone can get the hang of using the seeder in a minute or two. Production is only limited by the stamina of the operator and flow of components.

We tried a number of containers for holding the seed and ended up with a simple "Vee" trough. We also found that picking up seed a row at a time cuts down on skips. Bunching on tomatoes and peppers (depending on the conditions of the seed) can at times be a problem. A sharp rap of the seeder on the trough or edge of the bench will usually clear a bunch. We have tried using some dry talc, but is has not been too effective. Debris in some seed can be a problem, and from time to time, you have to clear the needles. First try reversing the air flow or have a fine needle handy.

Assembling the seeder is quite simple. The drawing shows all the pieces, components, and dimensions. Cut all pipe with a tube cutter. Drill the holes for the needles with a drill press. The more precise the rows and spacing, the better your results will be. Burn the plastic off each needle—quite simple to do. Do not glue the pieces of pipe.

You will spend about $10 to $12 in a hardware store (or with your friendly plumber), and about the same in a medical supply house. Don't use a home vacuum cleaner—you will burn it out (we did). A "Shop Vac" works very well. If you don't have a medical supply house handy, try a nearby hospital. Our local hospital was most helpful.

Small is beautiful

Adhara is alive and well, and is living proof of the validity of Schumacher's thesis "Small Is Beautiful" (a book worth reading regardless of your size). We have proven a small nursery can compete with the biggest and show an excellent return for the time and investment.

"

Paul King is the owner of Adhara Nursery, Gilmanton, New Hampshire.

Care and Handling

The latest research on holding and shipping plugs

December 1989

by Dave Koranski, Paul Karlovich and Abdalrahman Al-Hemaid

The how-tos of holding and shipping plugs, from detailed looks at tray size, temperature, length held and light. Based on the most recent research at Iowa State University using plug tray sizes of 200, 400, 600 and 800. The take home message is that the 800 plug from time of transplanting to time of flowering takes one to two weeks longer when compared to the 400 plug. The data below takes into account the time from sowing to flowering. However, the grower who buys plugs needs to know that it takes one to two weeks longer for flowering.

In the past five years we've discussed and researched many plug growing issues—water quality, nutrition, media, growth regulators and seed physiology. But, how do you hold and ship plugs? Increasingly, questions about these topics are being asked. How long can I hold my plugs? What is the effect of plug size on holding, growth and development? What cooler temperatures can I use, and for how long? If I ship in the dark, what are the effects? Will an 800 plug ship as well as a 500?

Hopper and Carlson wrote an article on the effects of plug holding at 60 degrees F. They studied five species grown in three cell sizes for four, six or eight weeks. Plants held for longer periods in the plug cell took longer to flower and produced poorer quality plants when grown at increasing plant densities. Do not hold plants in plug cells any longer than necessary was the recommendation.

Unfortunately, growers can't always plant at the correct time and shippers don't always have a buyer at the correct time. The results: many dumped plants and lost income.

The ability to hold plugs for several days to several weeks would be of great significance to growers and shippers. Plugs could possibly be placed in a reefer for one to two weeks, moved back to the greenhouse, acclimatized and shipped to the customer.

We conducted research to investigate the effect of temperature on holding plugs. Except for plants used in Experiment 2, all plugs were shipped from commercial growers prior to their use.

Tray size and holding: What's the relationship?

Petunia seedlings in 200, 406, 512 and 800 cell trays were held at 70 degrees F in the greenhouse for zero, one, two or three weeks. Plant height, days to flower and visual plug quality ratings were recorded.

Petunias seedlings were largest from 200 cell trays: 406 and 512 cell trays produced plants of similar height; and 800 cell trays produced the smallest plants (Figure 1). As holding time increased, plant size decreased for all plug tray sizes.

Flowering time was shortest in 800 cell trays and longest in 200 cell trays (Figure 2). Holding delayed flowering in all plug tray sizes. Three-week delays occurred in 200 plug trays held for three weeks.

Visual observations told the most important story (Figures 3 and 4). Reduced quality was observed as early as three days. By seven days, plant quality was obviously poorer than the original plants. Smaller cell sizes were more noticeably affected by holding than were larger cell sizes. After three weeks of holding, 800 cell plugs had poor quality.

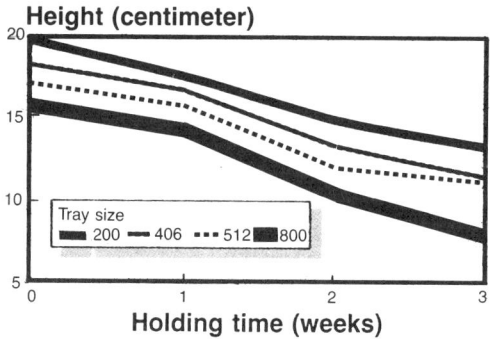

Height (centimeter)

Tray size
— 200 — 406 ··· 512 ▮800

Holding time (weeks)

Figure 1. Holding time and tray size definitely affect final plant height. During the holding period of three weeks, petunia plugs were largest in 200 cell trays and smallest in 800 cell trays. The 406 and 512 seedlings were similar in height. As holding time increased, plant size decreased in all plug trays.

The conclusion from this experiment is that petunias should be transplanted as quickly as possible to maintain the highest plant quality and to avoid flower delay.

Transplanting time

Petunia plugs (406s) were transplanted one week before normal, at the normal time, and one and two weeks later than normal. We observed growth and flowering times.

Dramatic effects for one week earlier transplanting were found. Plants showed increased growth and flowered two weeks earlier than plugs transplanted at the normal time. Planting at the normal time caused delays similar to those seen in Experiment 1 for plants held for one week.

Light and temperature interaction

Petunias, pansies, impatiens and vinca plugs in 406 trays were held at 40 degrees F, 50 degrees and 60 degrees F for one, two or three weeks in coolers with approximately 300 footcandles of light. After the holding period, the plugs were transplanted and grown in the greenhouse.

At 40 degrees F the plugs responded as follows:

● *Petunias* did well.

● *Pansies* did well except if moved abruptly to an 80-degree F greenhouse, where yellow bottom leaves resulted. Pansies should be acclimated overnight at a lower temperature (60 degrees F) before moving to the greenhouse environment. Pansies are the only plant having this problem.

● *Impatiens* chilling damage occurred after two weeks; damage included killed growing point and wilting. By two

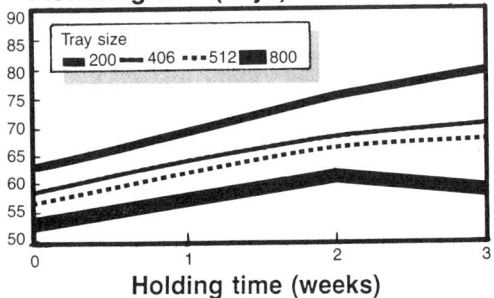

Flowering time (days)

Tray size
— 200 — 406 ··· 512 ▮800

Holding time (weeks)

Figure 2. Holding time and tray size affect flowering time—quickest flowering occurred in 800 trays. Holding delayed flowering in all plug trays.

weeks all impatiens were unacceptable for transplanting. Impatiens should be held for no more than one week at 40 degrees F.

● *Vinca* held well as long as plugs had a good root ball.

At 50 and 60 degrees F all cultivars did well. There was a little growth in the 60-degree F plugs.

Other than noted above, little or no difference in plant quality (as compared to immediate planting) was observed. A decrease in plant height, observed in plants held at 70 degrees F (Experiment 1), did not occur with these plugs, and flowering was not greatly delayed. Plants took three to four days longer to develop than normal.

Some of the petunia and pansy plugs in this experiment were left in the cooler for eight weeks. The light was turned off; the temperature was 60 degrees F to simulate shipping. Within three days, petunia plants that had been 1 inch tall grew to 4 to 5 inches and were ruined. The pansies also stretched, but more slowly and not as dramatically (Figure 5). At 40 degrees F, no petunia or pansy plant stretching occurred even after three weeks in the dark.

Holding plugs in the dark

Petunias and pansies in 406 plug trays were held in the dark at 40 degrees, 50 degrees or 60 degrees F for one, two, three, four or five days prior to transplanting.

At 40 and 50 degrees F no problems were observed. But at 60 degrees F, the petunias were ruined and the pansies yellowed and stretched.

Pansy rooting out was observed comparing 40 degrees versus 60 degrees F. The 40 degrees F plants took one day longer to root out. Other species have not been studied.

Figure 3. After holding petunia plugs for three weeks in the greenhouse, there is a noticeable size difference between 200- and 800-cell seedlings.

Holding plugs: the bottom line

Short-term holding at 70 degrees F is not advisable. Plugs can be held at 40, 50 or 60 degrees F for up to three weeks in the light.

Our recommended procedure for holding plugs:
- Light at 300 footcandles.
- Have a good root ball.
- Maintain low medium EC.
- Place plugs in cooler with dry plug medium.
- Water plugs only when needed.
- Fertilize at low rate.
- Treat plants with a fungicide if in the cooler for two or three weeks.
- Add three to four days more than normal to production schedules when transplanting.
- Acclimatize pansies before moving to a warm greenhouse.
- Hold impatiens no longer than two weeks at 40 degrees F.

Figure 4. Reduced plant quality is apparent as early as three days after holding plugs in the greenhouse. Plant size decreased and flowering delayed with increased holding time in a 200-cell tray.

Shipping plugs

One of the major shipping problems is shipping lush plugs. The result: Plugs arrive with curled and cupped leaves and brown tips. To solve this problem, tone plugs for one week prior to shipping by:

- Drying plants down.
- Lowering light intensity.
- Lowering temperature.
- Lowering fertility.
- Using growth regulators.

Most plugs will ship well at 45 degrees F. Pansies and impatiens need to be acclimated once received or yellow, cupped leaves and poor rooting out may occur.

Conclusion

The implications of this research are promising. Holding plugs at reduced temperatures in the light appears to have no detrimental effects in most cases on the species we studied. Holding plugs in the dark for extended periods is not recommended. Held plugs take three to four days longer to finish. Dark storage for short periods of time, such as the amount of time needed to ship plugs, was not detrimental to the plugs as long as they were toned and shipped at 40 to 50 degrees F. Some plant species, such as pansy and impatiens, need to be acclimated once they are received.

The dramatic increase in growth and reduced flowering time of petunias transplanted one week earlier than normal suggest that we may be able to further reduce production time by transplanting earlier. The implications of these results on holding and shipping need to be researched.

Future work will include more detailed studies on the effects of transplanting time and tray size on root and shoot development and flowering after holding and shipping.

"

David Koranski is a professor of Floriculture, Paul Karlovich is a recent Ph.D. graduate and Abdalrahman Al-Ilemaid is a master's candidate, Iowa State University, Ames.

Culture notes

January 1988

by Debbie Hamrick

The care and handling of bought-in bedding plant plugs

Plugs have become a major part of spring for thousands of growers. As many as one-half of the 3 billion-plus plants grown for the bedding plant industry in the United States are estimated to be started from plugs.

Some growers have gone "100 percent" to plugs for their bedding plant crops. Others hang onto sowing seed in open flats and transplanting bareroot. But most growers, whether they're growing their own plugs or still sowing in open flats, are buying in plugs. Why?

First, growers buying in plugs eliminate the hassles of sowing their own seed and having to run greenhouses for that purpose during the winter months. For example, why bother with such a long crop as begonia seedlings when you can buy them in?

Second, plugs are easy and fast to transplant so crews work quickly and efficiently. By working together with suppliers, plugs may be scheduled to arrive to keep work crews going steady, avoiding the mad dash in dealing with seedlings that always seem to be ready to transplant all at once.

Third, because plugs are substantial plants when they are planted, they take off and finish faster, so growers can turn their greenhouse space more often. Turning each square foot more puts more money on the bottom line.

And lastly, plug-grown bedding plant crops are uniform.

The benefits of growing bedding plants from plugs are major, but for the grower buying in plugs they are only realized when plugs are handled properly. Plugs are expensive—up to 6 cents each. Protect the investment and realize the benefits through handling the product properly when it comes through the door.

Inspect upon arrival

Inspect the shipment upon arrival. Report problems to the place of purchase immediately. (Dropped leaves and leaf yellowing may be signs of heat buildup during transit.) Water trays with clear water if needed. Remember to check plug trays for water often—maybe as much as three to four times daily—their small soil balls dry out quickly. And be sure to provide irrigation drainage away from the bottom of plug trays.

Acclimatize plants and give them time to recover from shipping by placing trays in a house maintained at 65 F for one to two days. Avoid direct sunlight for the first couple of days.

Transplanting

The entire bedding plant plug system is meant for immediate transplanting. Holding these actively growing plants slows their growth process and adds time to production. Plugs should be transplanted within three to five days of arrival; this ensures that plugs will remain healthy and actively growing. When plug trays are held longer than one week, you risk loosing actively growing root hairs. Plants may also become root bound. And root bound plants do not take off as quickly, increasing total crop time.

If plugs must be held, be careful of temperatures. The goal is to keep roots healthy and actively growing. Soil temperatures should not drop below 62 F to 65 F. *Temperatures below 60 F for more than one week will delay flowering.*

The best way to avoid plug tray backup and holding plugs during the busy season is to develop a cropping schedule and work closely with plug suppliers to realistically time plug shipments.

Before transplanting, water plug trays a couple of hours beforehand—this makes them easier to dislodge. To avoid damaging plants, use a mechanical plug extractor (available from several equipment companies) or a small cylindrical object such as a golf tee to manually dislodge plants from their cavities. Do not separate multiple seedlings in the same plug.

Media: For optimum root development, transplant into a mix with 20 percent air porosity—coarse peat or calcined clay may be added to the mix to increase air porosity. Media pH should be 5.8 to 6.0. The initial nutrient charge of the media should be low, below 100.

Other factors which may inhibit root growth include high bicarbonates or a high sodium absorption ration (greater than 2) in the irrigation water.

Fertilizer and other chemicals: Before beginning any fertilizer or other chemical program, allow time for roots to break out of the plug root ball, usually in three to four days.

Examine the root systems of check plants to find out when roots have penetrated media outside of the soil ball—when they've broken the soil ball, generally, it's time to fertilize.

Last year, many growers, especially in the Southeast, had problems with vinca plugs. When the plugs were transplanted, they never greened up, never penetrated the media outside the root ball. This problem has been attributed in part to fungicide treatments (drench was used), in part to high salts in the media and in part to fertilizing too soon. The moral: Plugs are small plants and the period of days following transplanting is critical and is not the time to apply lots of chemicals.

Other crops known to be sensitive to chemicals during this critical phase of development include impatiens, Non-Stop begonias and salvias. If you are concerned for any reason, go easy on fertilizer—half-strength—at first. But always make sure media is low in nutrient charge and avoid any pesticides until roots are established.

Temperature: To get plants off to a good start, maintain night temperatures between 68 F and 70 F. Temperatures may be lowered to 60 F to 62 F nights once flowers have been initiated.

Once plugs are off and growing in the flat, pot or basket, treat them as a regular bedding plant crop.

"

Debbie Hamrick is editor of GrowerTalks *magazine.*

Insects and Diseases

Identifying the most serious problem we face today—TSWV

July 1989

by Russell Miller

Visual symptoms will not provide conclusive evidence of TSWV, but they are a good cause for concern.

The last thing this industry needs is panic over a problem few growers have experience dealing with, say plant pathologists and entomologists in the United States and Canada. Tomato spotted wilt virus, however, is widespread and a serious threat to all commercial greenhouse growers—many who may not be aware that they may be dealing with TSWV-infected plants in their greenhouses right now.

A worst-case scenario guesstimate by plant pathologists and entomologists is that at least 40 to 50 percent of cutting-propagated impatiens, including New Guinea impatiens, in greenhouses this year are infected with TSWV.

Most plant pathologists and entomologists call TSWV the No. 1 problem the greenhouse industry faces today. All growers should become very familiar with identifying visual symptoms of TSWV, which vary according to plant species and the strain or mutation of TSWV involved. In addition, there are different "sets" of visual symptoms on several plant species, and in some cases TSWV visual symptoms

TSWV symptoms of gloxinia: On young plants—stunting, necrotic line and ring patterns, terminal bud necrosis and brown or collapsed new growth, similar to *Phytopthora parasitica* infection. On older plants—chlorotic to necrotic rings, yellow or brown ringspots, oak leaf patterns, midrib browning or line patterns on leaves (appearing more obvious on the lower leaf surface than the upper surface), and distorted flowers with distinct line and ring spot patterns of color break (white in normally colored petals). Gloxinias are one of the few crops that may actually die from the infection alone. Young gloxinias are highly susceptible and are apt to show visual symptoms more readily than older gloxinias. The symptoms are likely most noticeable shortly after the plant becomes infected. Surviving gloxinias may become symptomless even though they still carry the virus. *(Top photo by Ron Jones, North Carolina State University; bottom photo by Larry Barnes, Texas A&M University.)*

Some New Guinea impatiens cultivars are highly susceptible and develop more obvious symptoms than other cultivars. Darker cultivars are more frequently infected.

TSWV symptoms of New Guinea impatiens: Stunting, black ringspots or dead areas (often on the petiole end of the leaf), small twisted leaves, some terminal die-back and necrotic leaf spots. *(Bottom photo by Larry Barnes, Texas A&M University. All other photos not credited in this article were submitted by a Midwest grower.)*

may not closely resemble the "classic" symptoms shown in photographs.

TSWV symptoms within the same plant species may vary depending on the age of the plant, when it became infected, nutritional levels and environmental conditions, especially temperature. In many cases TSWV symptoms may be easily confused with symptoms induced by other viruses, fungal or bacterial pathogens, nutritional disorders or spray damage. Generally, TSWV symptoms include stunting, ringspots, dark purple-brown sunken spots, stem browning, flower breaking and death.

Key points about TSWV

• **The predominant vector of TSWV is the western flower thrips** (*Frankliniella occidentalis*). The western flower thrips is the most efficient vector in transmitting the virus to plant material, the most difficult pest to control in greenhouses, and the predominant thrips species encountered in greenhouses. Other thrips species, however, can also be TSWV vectors. Weeds inside and outside greenhouses should be removed, as weeds may also serve as host plants.

Both larvae and adult thrips can transmit the virus to plant material, and usually thrips larvae acquire the virus from TSWV-infected plants. Once a thrips larva picks up the virus, it remains infected for its adult life—30 to 45 days. Transmission of the TSWV virus by thrips may be sporadic. Monitor thrips with blue or yellow sticky cards and rotate chemical classes for control every one to two months.

• **TSWV is an industrywide problem** and all growers share the responsibility for controlling and preventing the spread of TSWV. The large commercial plant suppliers who have taken steps in the past to ensure that growers receive clean plant material—as well as any grower who can reduce the number of western flower thrips in greenhouses—offer the industry the most promising means of reducing the spread of TSWV through virus-infected plant material and virus-carrying western flower thrips.

• **Few greenhouse crops are safe from TSWV.** Outside of geraniums, poinsettias and roses, most other greenhouse crops are quite susceptible to TSWV. The host list includes nearly 200 plant species in some 40 plant families. Geraniums are listed as a TSWV host, but crops are not currently being harmed by the TSWV strains that are circulating in the greenhouse industry.

TSWV symptoms of double-flowered impatiens: Stunting, leaf midrib browning, necrotic spots and ring patterns in leaves, target-shaped or black ringspots. Impatiens are highly susceptible. Seed impatiens should be physically separated from cutting impatiens. *(Top photo by Gail Ruhl, Purdue University; others by Margery Daughtrey, Cornell University.)*

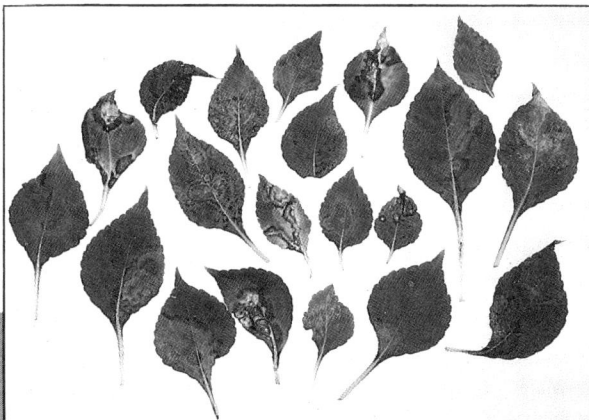

The most common plants being submitted for TSWV testing are cutting impatiens, *Impatiens wallerana*, New Guinea impatiens, chrysanthemum and gloxinia. These crops are common in greenhouses, which at least partly explains why these plants are being tested for TSWV often. They also are more susceptible to TSWV and show TSWV visual symptoms more readily than other plant species. These plants, however, are not the only plants growers should pay careful attention to. All bedding plants from seed are at high risk for TSWV infection.

• **Viruses rarely kill plants,** but usually weaken them, causing poor growth and susceptibility to pathogens that can kill them. TSWV *is* capable of killing highly susceptible species or cultivars under certain circumstances. There are a few general observations: TSWV-infected plants may never show visual symptoms; young plants are likely to display visual symptoms more than older plants; some infected plants may partially recover; and plants from cuttings probably show TSWV symptoms more readily than the infected stock plants they came from.

A TSWV-infected plant may appear to have a healthy, vigorous root system, yet above the soil line the plant may have severe TSWV symptoms, or even be on the verge of death. All stages of plant growth are capable of acquiring TSWV from infected thrips.

• **TSWV is not a new problem nor a U.S. problem.** TSWV is showing up in greenhouses in the United Kingdom. It's a devastating problem in tobacco fields in Rhodesia and a problem in peanut and tobacco fields throughout the

The impatiens strain of TSWV on impatiens, as shown in these three photographs. *(Photos by Brian Whipker, Purdue University.)*

A great step forward in accurately identifying TSWV is a new antisera against the impatiens strain of TSWV, which, incidently, is not limited to impatiens. The antisera, developed at North Carolina State University, identifies this strain that was previously being missed in other tests.

'GrowerTalks'® magazine

New Subscription Order Form
—card expires January 1992—

_____ 3 year **$50**
($70 Canada/Mexico, $146 Other)

_____ 2 year **$35**
($47 Canada/Mexico, $99 Other)

_____ 1 year **$19**
($25 Canada/Mexico, $51 Other)

Published monthly   **U.S. Currency only!**

Payment must be enclosed. *Please note any changes to your Name/Address below.*

☐ Check/money order   Charge my: ☐ VISA ☐ MASTERCARD ☐ AMERICAN EXPRESS

Account number: ☐☐☐☐☐☐☐☐☐☐☐☐☐☐☐☐

Expiration date: ☐☐ – ☐☐   Your signature (required) _____

Cardholder's Name (please print) _____

Title _____

Company _____

Address _____

City _____ State _____ Zip _____

Phone ( ) _____

Signature _____

Date _____

---

1. I am a grower ☐ Yes ☐ No   PLUGS 590

2. Crops grown (check all that apply)
☐ a. Bedding plants
☐ b. Pot plants
☐ c. Foliage
☐ d. Cut flowers
☐ e. Vegetables
☐ f. Plugs
☐ g. Perennials
☐ x. Other_____
☐ y. None of the above

3. Growing area under cover (check only one)
☐ 1. 1-9,999 sq. ft.
☐ 2. 10,000-24,999 sq. ft.
☐ 3. 25,000-49,999 sq. ft.
☐ 4. 50,000-99,999 sq. ft.
☐ 5. 100,000-249,999 sq. ft.
☐ 6. 250,000 sq. ft. and over
☐ 7. Open field_____acres

4. Nature of business (check all that apply)
☐ a. Retail/Wholesale
☐ b. Retail
☐ c. Wholesale
☐ d. Landscape
☐ e. Garden center
☐ f. Nursery
☐ g. Retail Florist
☐ h. Institutional/Educational
☐ i. Government/Extension
☐ j. Manufacturer
☐ k. Supplier/Distributor
☐ x. Other _____
☐ y. None of the above

5. Title (check only one)
☐ a. Owner/President
☐ b. General Manager
☐ c. Production Manager
☐ d. Head Grower
☐ e. Production Assistant
☐ f. Director/Superintendent/Mgr.
☐ l. Vice President
☐ m. Marketing Manager
☐ n. Sales Manager
☐ o. Office Manager
☐ r. Educator/Researcher
☐ s. Student
☐ x. Other _____

Do you want to receive/continue to receive *GrowerTalks* magazine? ☐ Yes ☐ No

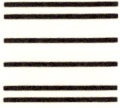

BUSINESS REPLY MAIL

FIRST CLASS   PERMIT NO. 447   GENEVA, IL

GrowerTalks magazine
Attn: Circulation
Post Office Box 532
Geneva, IL 60134-0532

TSWV symptoms on seed impatiens that acquired the virus from thrips: plant distortion, leaf spotting and stem discoloration.

TSWV symptoms of cineraria: Flower distortion, mottling or yellow ringspots on leaves, necrotic spotting on lower leaves and chlorotic spotting on upper leaves, dark spots on leaf blades and dark streaks on leaf petioles. May look similar to phytotoxic spray injury. TSWV may be carried within cineraria seed. See *GrowerTalks* magazine, March 1988, page 102, for more photos. *(Photo by the University of Guelph.)*

Gulf states. TSWV is worldwide and growing in proportion with the spread of the western flower thrips.

• **If you suspect TSWV-infection,** you'll need to submit plant samples to a diagnostic laboratory for confirmation and begin a management program, which includes immediately destroying infected plants and controlling

Portulaca infected with TSWV.

TSWV symptoms of begonias: Dark brown lines along leaf veins, severe necrosis of leaf veins, yellow or bronze ringspots, blotches, mottling and delicate follar necrotic ring spots. The effects on Rieger begonias may appear on only one or two leaves. Color breaks in the form of white rings may show up in red and pink flowers. May be confused as *Xanthomonas campestris* pv. *begoniae*. (Photo by Ron Jones, North Carolina State University.)

the western flower thrips. Be prepared to submit at least two plant samples at two different times or to two or more different labs. TSWV is difficult to confirm in the lab, and false negative test results occur. Negative test results (virus not present) do not always mean a plant is free from TSWV.

Several U.S. and Canadian universities have TSWV diagnostic laboratories. Contact university horticultural departments for more information. There are at least two nationwide TSWV testing laboratories accepting plant material from growers: **Agdia Inc.**, 30380 County Road 6, Elkhart, Indiana 46514, (219) 255-2817; and **Yoder Brothers Inc.**, (Jane Trolinger), P.O. Box 68, Alva, Florida 33920, (813) 728-2535.

• **For more information on TSWV** see the following articles in *GrowerTalks* magazine: "Growing ideas," June and July 1989; "Working the bugs out," December 1988 and February 1989; "Dealing with tomato spotted wilt virus," March 1988; "An integrated approach to preventing WFT and TSWV in the greenhouse" and "They're invisible and incurable. Are viruses invincible, too?" March 1989.

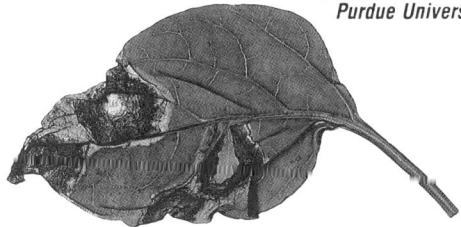

TSWV symptoms on pepper. (Photo by Gail Ruhl, Purdue University.)

A valuable reference for growers is the publication, "Tomato Spotted Wilt Virus: A Serious Threat to Greenhouse Crops," by Charles Powell, Stephen Nameth and Richard Lindquist, Departments of Plant Pathology and Entomology, The Ohio State University. Single copies are available free by writing to Extension Plant Pathology, The Ohio State University, 2021 Coffey Road, Columbus, Ohio 43210.

"

Russell Miller is features editor of GrowerTalks *magazine.*

Thanks to the following for contributing photographs or technical information for this article: Larry Barnes, Texas A & M University; Joe Begley and Jane Trolinger, Yoder Brothers Inc.; John Cho, University of Hawaii; Margery Daughtrey, Cornell University; Gail Ruhl and Brian Whipker, Purdue University; Jim Matteoni, Agriculture Canada, Vineland Research Station, Ontario; Jim Moyer and Ron Jones, North Carolina State University; Steve Nameth, The Ohio State University; and Chet Sutula, Agdia Inc.

Working the bugs out

June 1989

by Mark E. Ascerno

Control fungus gnats during production.

F ungus gnats continue to plague many growers. Correct identification is essential for good control as fungus gnats can be easily confused with other insects.

Identifying adults

Fungus gnat adults are less than ⅛-inch long, fragile, gray to black flies with relatively long legs. Because they are true flies, they have only one pair of wings.

The presence of a distinctive "Y"-shaped vein in each wing distinguishes fungus gnats from shorefly adults. This characteristic is best seen under magnification. Not requiring magnification and easier to see are antennae on the fungus gnat that are noticeably longer than the head.

The adults do not cause any damage to plants, but when abundant, they can become a nuisance. Customer complaints, particularly in retail operations and where plant materials are being sold in supermarkets, create a lot of the pressure to control fungus gnats during plant production.

Fungus gnat larvae damage roots

As is true of all flies, fungus gnats have an immature stage that does not look at all like the adult. The larvae are worm-like, translucent to opaque white and have a distinctive shiny black head capsule.

While the adults don't directly damage plants, the larvae do. They normally develop in fungi-infested soils and in decaying plant material. Once established, the larvae feed on small roots and root hairs. In some cases, larvae may extend their damage by tunneling in succulent stems. For example, they have been found tunneling poinsettia and geranium stems, particularly when root rot was present.

Fungus gnat adults are characterized by a Y-shaped vein in each wing and antennae that are longer than the head.

High humidities commonly associated with plug crops and the fact that fungus gnats produce their greatest damage on tender plants make plugs more susceptible than conventionally produced crops.

Start control under the bench

Proper sanitation is important for reducing fungus gnat populations. Growers who allow weeds to remain beneath benches are asking for problems. The weeds help to trap moisture and contribute decaying organic material that serves as food for the fungus gnat larvae. For most growers, fungus gnats get their start under the bench before attacking the crop.

Several approaches can be used to make the under-bench environment inhospitable for fungus gnats. Keeping the area free of weeds and plant debris will reduce the amount of food readily available. This can be accomplished by hand weeding and clean-up or by the use of proper herbicides.

Some growers use other techniques to eliminate weeds and prevent fungus gnat development. One approach employs copper sulfate at 1 pound per 1 gallon of water. The mix is sprayed on all under-bench soil. A three-month residual is claimed at this rate.

While growers using this approach feel it's effective, I have been unable to find any research data to verify these claims. Be careful not to allow the mixture to contact your crop as it will burn living plant parts.

Insecticides can also be used under the bench to control fungus gnat larvae. Diazinon, oxamyl, bendiocarb (Turcam, Dycarb) or Gnatrol should be effective for this purpose.

In-crop control can't be ignored

Even with good beneath-bench control, a grower may find that in-crop control is needed. Diazinon (Knox-Out), oxamyl, Gnatrol, and bendiocarb (Turcam, Dycarb) have been reasonably good at controlling fungus gnat larvae. Since the majority of larvae feed close to the soil surface, light drenching is usually effective.

In addition to the larval stage, the adult fungus gnat may also need to be controlled, particularly if the population has gotten out of hand. Resmethrin in aerosol form usually knocks out adult populations.

Monitor with yellow cards

Monitoring the fungus gnat population is also important in a good control program. The adult stage can be trapped on yellow sticky cards hung just above the crop surface. Regularly check the cards and record the number of gnats found. A population that is increasing will require greater control efforts, while one that is decreasing suggests your techniques are working.

Don't forget root rot control

Finally, since fungus gnat larvae do best when fungi are present, controlling soil fungi, particularly root rots, and using the appropriate fungicide will help reduce fungus gnat problems. Check your local extension service for recommendations in your area.

"

Mark E. Ascerno is professor and extension entomologist at the University of Minnesota, St. Paul.

Understanding and controlling black root rot disease

March 1989

by Chuck Powell

Plug-produced pansies and vinca are under fire from thielaviopsis—black root rot. To control this devastating pathogen, you need to clean up, get soluble salts down, drench routinely with fungicides and pamper your plants with a stress-free environment.

Black root rot is one of those classical, troublesome root diseases that was first described on greenhouse crops many years ago. The pathogen, *Thielaviopsis basicola*, is very widespread, having been observed and reported on hundreds of kinds of plants including cotton, tobacco, poinsettia, geraniums and fuchsia.

The fungus is a soil-borne organism capable of living in soils as a saprophyte (without causing disease). The organism can survive in soil and dust for years via a tiny, thick-walled spore called a chlamydospore.

Whereas black root rot caused widespread losses in poinsettia crops 20 years ago, it has of late been rarely seen in poinsettias. Such is not the case in plug-produced bedding plant crops.

Growers throughout the United States have logged losses in plug-produced pansies and vinca over the last couple of years. In this article, I'll show how to diagnose black root rot, try to explain some reasons for the new emergence of this old disease, and outline disease management practices.

Diagnosing black root rot

Thielaviopsis can cause sickness in plants in several ways. According to research results, the fungus can act as a pathogen when it lives close to but not within the host plant's roots. The fungus produces a toxin in such cases, which causes plant stunting, leaf yellowing and poor growth. Under other conditions, thielaviopsis can infect plant roots and lower stems, producing black cankers of varying sizes. Cankered roots eventually wither and die.

The black root cankers can be seen relatively easily by washing suspect tissue free of growing media particles and then viewing them carefully with a hand lens. Infested roots will be without root hairs. They will often be wet or mushy looking in the early stages of infection. The canker tissue is usually filled with barrel shaped chlamydospores, which can be easily seen under the microscope. If you are really good, you can see these spores with a 20 X hand lens.

As root cankers develop, the plant gets sicker and sicker, eventually dying. Black root rot causes rather general symptoms on leaves and shoots. It can only be diagnosed properly by finding cankered roots or stems and then confirming the presence of the chlamydospores with a microscope.

Infected plants, media and dust spread disease

As you might guess, thielaviopsis can spread between greenhouses or between crops within a greenhouse in many ways. Long distance spread between greenhouses recently seems to be occurring primarily via the movement of infested—but not necessarily parasitized—plant material. Such plants may look perfectly healthy. It may be impossible to detect the presence of the pathogen on such plants via standard lab procedures. This situation presents a potentially serious problem that all of us must address.

Thielaviopsis also has the potential to enter a greenhouse via wind-blown dust or with infested growing media. Black root rot can be found on many field crops, especially those in the Sun Belt states, such as cotton and tobacco. We do not know if the fungal isolates or strains we have recently encountered on vincas and pansies are the same as those found on agronomic crops like cotton and tobacco. Until research is completed on this strain situation, we must presume such pathogen movement into greenhouses can occur.

Seed crops for many of these affected plants are field grown. Their seed are sometimes harvested in ways that could result in plant debris or dust getting mixed in with the seed. Thielaviopsis chlamydospores could survive such treatment. The bottom line, however, is that—to date—no thielaviopsis-infested seed lots have been detected in the bedding plant industry. Pathology labs in Texas, Georgia, California, here in Ohio and elsewhere are continuing to research this possibility.

Thielaviopsis invades and moves in to stay

Once thielaviopsis gets inside a greenhouse, many situations result in the fungus becoming a long term resident or an endemic problem. Its broad host range and saprophytic nature mean that it may continue to grow and survive on many plants in the greenhouse that may appear symptomless or nearly so. As I mentioned above, we don't really know precisely the host range of the isolates that have recently been causing the problem on vinca and pansies.

Chlamydospores are produced on cankered host tissue in tremendous numbers. These spores can be splashed about or blown about as dust in the air. They can ride along with crop debris or in bits of infested growing media throughout the greenhouse, into the headhouse, into transport trucks, etc. They will be present on flats, pots or trays that may be brought into a work area for reuse.

In addition, the pathogen produces a second type of spore, very tiny and very well adapted to short distance spread on the root system of a single plant. These spores are also well adapted to water splash spread. It may be the spread of these small spores that allows the disease to develop so quickly once it gets started on a host plant. All in all, thielaviopsis is well suited to become a permanent, though unwelcome, guest in your greenhouse!

Managing black root rot

Black root rot is manageable! Many integrated procedures—taken together—will result in a very high level of disease control. This disease, like so many, requires an integrated program of holistically oriented health management practices. Practicing one of the following procedures without the others will not be cost effective.

• **Stress management is crucial.** Almost any factor that stresses host plants will lead to further parasitic development of the pathogen. On the other hand, correcting plant stress often results in such effective pathogen impedance that infested plants will "outgrow" the disease and eventually recover.

Most growers who had black root rot in the fall pansy crop this year experienced sizable losses in September when it was hot, but minor losses in November as the weather cooled. The fungus could be seen on November plants, but there was little plant sickness evident.

It is difficult to say precisely which stresses are responsible for the current black root rot outbreaks we are seeing. Temperature problems are suspected. Crops grown too cool or too warm are often seen to be badly diseased.

Nutritional imbalances are also frequently associated with disease expression. Excesses of ammonia nitrogen appear to be particularly troublesome for pansy growers. Growing media pH levels between 5.5 and 6.0 are also conducive to disease development. More research is needed to be more precise.

● **Sanitation routs disease problem.** *Thielaviopsis basicola* is well adapted to persistence in the greenhouse. Once it is known to be present in your greenhouse, you must work hard at improving sanitation to get rid of it or keep it in bounds.

The needed techniques are classical, but not complicated. They are not necessarily cheap or easy to do. Do not reuse containers or trays. Do not create dust, especially when the dust can settle on piles of growing media, on new pots or trays, or on healthy crops growing nearby. Sanitize benches, floors, etc. with a Q-salt such as Physan 20. Use one-half ounce per gallon of the 20 percent active ingredient product and soak for 10 minutes or longer.

● **Chemicals help eliminate pathogen.** Benzimidazole fungicides (Benlate DF or Cleary's 3336) are the only ones currently labeled and known to be effective against thielaviopsis. Routine, monthly drenches at labeled rates will help protect your crops. With bedding plants, you must have the fungicide present at seeding. Continue on a rigorous monthly schedule throughout transplanting and finishing the crop.

Most growers will also be using a second fungicide in a drench program for water mold control (pythium and phytophthora). Evidence for the detrimental effects of metalaxyl (Subdue) is present. Many growers have had worse black root rot problems when they use metalaxyl. Recent research at Texas A&M confirms this.

The mechanism may be the tendency of metalaxyl to inhibit the bacterial conversion of ammonia to nitrate (see above under stress and ammonia nitrogen). Until this effect is more thoroughly researched and understood, I am suggesting that growers with black root rot problems use propamocarb (Banol) or phosethyl-A1 (Aliette) for water mold protection.

You can beat black root rot

Black root rot has been around for years. In spite of its recent flair up, I believe it to be a manageable disease. Employ good, holistic and integrated health management procedures. If you suspect black root rot, confirm the diagnosis and initiate further activity as outlined above.

”

Chuck Powell is professor of plant pathology, Ohio State University, Columbus.

Working the bugs out

January 1988

by Mark E. Ascerno

"Plug tissue is far more susceptible to damage by pesticides."

Is insect control in plug crops different than control in more conventionally produced crops? To my knowledge, no one has looked at this question experimentally. Although the pests associated with plug crops run the range of common greenhouse problems—fungus gnats, spider mites, thrips, leafminers, whiteflies—extra care may be needed when selecting the pesticides that will be used for control.

Plugs result in young, tender and succulent plants growing in a small amount of media. Since plug tissue is far more susceptible to damage by pesticides, we can expect phytotoxicity to be more of a problem in plug production. In addition, application rates for soil-applied systemic insecticides are based on a relatively large soil-to-root ratio. The small soil-to-plant ratio in plug production requires *precise* application rates. A slight over-calculation may result in an over-concentration of chemical, causing a severe phytotoxic reaction. For this reason, it is best to avoid the use of systemic insecticides until you have worked out the proper rates through small scale trials on a per crop/variety basis.

The chart lists crops commonly grown as plugs, and *known* phytotoxicities of common greenhouse pesticides when conventionally grown. This list is only advisory, but since plugs are more sensitive it should indicate potential problems when the crops are grown as plugs. I emphasize that any chemical may cause phytotoxicity when applied under unfavorable environmental conditions. Always use caution by testing a few plants before applying any

Pesticides for annual plugs

Crop	Pesticide	Phytoxicity remarks
Marigold	Pentac	Certain varieties
Impatiens	Insecticidal soap	Certain varieties
Petunias	Insecticidal soap	Certain varieties
	Malathion	Certain varieties
Begonias	Orthene	Certain varieties
	Temik	Certain varieties
	Systox	Floral and plant parts
	Diazinon 50WP	Floral and plant parts
	Kelthane	Certain varieties
	Cygon	Plant parts
	Malathion	Floral parts and certain varieties
	Meta-Systox-R	Certain varieties
	Parathion	Floral parts and certain varieties
	Sulfotepp	Floral parts and certain varieties
	Vapona	Floral parts and certain varieties
Geraniums	Enstar 5E	Floral parts
	Lindane	Floral parts
	Meta-Systox-R	Floral parts
	Parathion	Plant parts
	Sulfotepp	Plant parts
	Vapona	Plant parts
	Tetradifon + Sulfotepp	Floral parts
Vinca	Dithio	Floral parts

From: Nancy L. Olson and Mark E. Ascerno, 1985. Pesticide table of phytotoxicity from pesticide applications to commercial greenhouse crops. university of Minnesota Agricultural Extension Service Miscellaneous Publication AG-MI-2659. It is available as a laminated wall chart containing phytoxicity information on 36 insecticides covering 72 common flower, bedding and foliage crops. Order from the Distribution Center, Coffey Hall, University of Minnesota, St. Paul, Minne

pesticide to a whole crop.

Fungus gnats and plugs

Of the common greenhouse pests, fungus gnats probably occur more frequently and create more problems in plugs versus conventional production. This is related to the high humidities commonly associated with plug crops and the fact that fungus gnats produce their greatest damage on tender plants as the gnat larvae feed on small roots and root hairs.. Because of this, most plug growers find themselves dealing with fungus gnats. Briefly, good fungus gnat control involves elimination of beneath-the-bench fungus gnat populations and in-crop insecticide applications.

"

Mark E. Ascerno is professor and extension entomologist at the University of Minnesota, St. Paul.

Plug Grower Profiles

Swift Greenhouse customizes plug range for perennial comfort

December 1989

by Russell Miller

When the Swift family decided on a new greenhouse, their hunt led them to a comparatively inexpensive poly range custom-built for growing 175 different perennials.

Perennials are 65 percent of greenhouse crop production at Swift Greenhouse, a third generation family business in Gilman, Iowa. During the 1970s, bedding plants and potted crops were the primary crops grown at Swift Greenhouse, but during the early 1980s, perennials caught on fast. They now have moved out of pot plant production in favor of perennials, both plugs and finished crops.

The Swift family began growing plugs for their own use in 1980. Today, plug growing accounts for one-third of the total crop production. The remainder is finished bedding plants. Swift Greenhouse now sells 8 million plugs a year through brokers nationwide.

Perennials have become a key crop in Swift's 100,000 square feet of greenhouse production area. They grow 175 different varieties. "Today, three out of every four plug flats we now sell are perennials," says Scott Swift.

New range custom-built for plugs

Swift's tore down several greenhouses in 1987 to accommodate a new 22,112-square-foot Van Wingerden triple-truss greenhouse, custom-built to fit Swift's production needs. The greenhouse has an A-frame design with 11- to 12-foot gutters for good air circulation, as well as to funnel condensation drips to the gutters. The greenhouse is also equipped with six 48-inch Aerovent cooling fans. In December 1988, they added Van Wingerden shade-heat retention curtains and eight Jaderloon horizontal air fans.

The 200-foot-long greenhouse features a 36-foot clear span divided into three 12-foot bays. Rolling benches from Minnesota Distributing and Manufacturing are used throughout the range. Two of the three bays are used for Stages 1 and 2 plug production, and the third bay is used for holding plugs prior to shipping. Irrigation is with automatic watering booms from Integrated Tech Systems.

"It's not your typical poly house, it's specially built for plug growing," Scott says. "We looked around before deciding on this Van Wingerden plug range. We talked to a lot of greenhouse manufacturers and we talked to Dave Koranski, who, fortunately for us, lives only an hour away. We wanted a structure that would accommodate automation in the future. So we looked for awidespan house that could also be divided into zones with the flexibility to accommodate crop production changes in the future.

"We looked at glass and Exolite before deciding on poly. We stayed away from glass because we don't grow crops 12 months a year—we don't do poinsettias or pot mums. We grow cold treatment perennials in the winter, then switch to plugs, and then finished bedding plants as the final crop in the spring. Also, the payback period for a glass house would be increased and heating in the winter would be more of a problem. Exolite is nice, but expensive and hard to erect if you are doing the construction yourself."

The new poly covered house features 85 percent space utilization for production, compared to 70 percent utilization in older house.

"With a 15 percent increase in bench space utilization, there's not going to be as many extra years in the payback period for this new house, compared to other ranges," Scott says.

The north wall of the range is Reflectex corrugated steel; the south wall is polycarbonate. For the north wall, they wanted material that would provide a good seal and a high R value.

"When we built the greenhouse we were looking at an insulated north wall, but we were being quoted at $2 to $3 per square foot," Scott says. "So we put up treated lumber covered with Reflectex on both sides. The cost of the material was 37 cents per square foot which is phenomenal. It's waterproof and reflective, and has an R value of 9. We also put Reflectex on the inside walls of our germination chamber."

Greenhouse split into eight zones

They split up the new greenhouse into eight heating and bench zones to produce both plugs and finished crops. There are nine benches per zone, for a total of 72 benches. They chose to install wide, flattened galvanized steel benches: Each bench is 10 feet, 8 inches wide by 16 feet, 4 inches long.

"The wider rolling benches are not as tippy as the common rolling benches," Scott says, "and plug trays don't catch when you are trying to move them on the flattened steel. The flattened steel benches, however, as they come, are not as strong as regular galvanized steel benches, so we had to reinforce them with cross members spaced at 16 inches."

Each bench holds 108 plug trays, six across and 18 deep. With the wide rolling benches, they can create 2-foot aisles between the benches. The main aisle lengthwise down the center of the greenhouse is 3 feet wide.

"We wanted to have wide aisles between the benches because we occasionally have more than one person working in a zone at a time," Scott says. "With wider aisles, two people can comfortably work together without bothering each other if they have to move in or out of the zone."

Each zone has its own water pump and thermostat for the fin tube heating, fed by a 4-inch main running down the center aisle. Scott says the only problem with having eight different heating zones in the greenhouse is that it's difficult to maintain more than a 5-degree temperature variation.

"We probably over-equipped the greenhouse by having eight heating zones," he says. "Eight zones is a lot for a small house, and it's expensive when you consider the pumps and materials needed for each zone, but we grow a lot of different plug crops and we need this flexibility. A lot of times we will have a crop that takes up only one bench, or four or five trays of different crops in varying growth stages on a bench. I don't know a way we could have done it differently and still maintain the flexibility we now have."

They also went with five Utica propane boilers for hot water heat: Each boiler produces 230,000 Btu.

"We went with five small units instead of two large units because of the increased operating efficiency of the smaller, multiple units," Scott says.

The boilers are in the head house, adjacent to the plug house, on 7-foot stands. On the stands, the boilers don't take up valuable head house space. As a backup, there are two Sebring forced air, fuel oil heaters.

Germination chambers are better

Three germination chambers are used at Swifts, each kept at certain temperatures for specific crops: 70 degrees F for perennials, 75 degrees F for annuals and 80 degrees F for annuals and waxed begonias. Each chamber measures 14 feet by 12 feet by 8 feet high and can be filled with up to six Cannon carts, or over 700 plug sheets.

"The chambers work out well," Scott says. "We can germinate as many as 2,000 flats every four to six days. For the major part, the germination chambers allow for more temperature control than germinating plugs on the greenhouse bench, and it's more economical because we save on bench space by germinating in the chambers."

Half the range is lighted by 72, 400-watt HID lights from P.L. Light Systems. About 350 footcandles of light is provided 18 hours a day; less depending on the availability of natural light.

The ITS booms are automated for each bench crop, which are marked by bar codes so that the booms automatically adjust the water applied to specific crops. Scott says they should have made their 2-inch incoming water lines larger to provide more water pressure in the plug range—3- or 4-inch lines would be more appropriate. "You can never really have enough water pressure in a plug range," he says.

"

Russell Miller is features editor of GrowerTalks *magazine.*

Wagner's state-of-the-art plug growing

December 1989

by Vic Ball

Skillfully automated and quality oriented, Wagner's produces plugs using a computer, carts and carefully controlled environments.

It is state of the art. Wagner's do about 2 acres of them at spring peak. Location: Minneapolis, Minnesota. The creators are the Wagner family: Rich and Nola are generation III, Scott and Ron are IV. Ron is production manager. We had a fascinating morning going through it all in mid-October.

Right up front, I've seen plug specialists 10 times bigger, and very good plugs—but none anywhere with a better quality orientation, more skillfully automated, and none that uses all available environmental controls, again more skillfully (computer-controlled)—to achieve their goal of top quality.

Let's start with the Dutch trays

All plug flats go right from the seeder to a 5- by 11-foot aluminum Dutch tray, (66 inches by 129 inches). And they stay right on it until ship out. These trays are the bench and they are basic to much of Ron's system. How so?

• **"We move the trays, often weekly** from one environment to another,"says Ron. "At first, higher temperature, fog, HID light. Then gradually we tone, harden them off. And the point is that we are able to create any desired environment in each of our greenhouses. And we move these trays from environment A to environment B—which is basic to producing that quality we are striving so hard for."

• **Labor costs** are also very important. "The cost to move all these trays back and forth is near zero," says Tom. "Trays are again the key to low-cost plant moving around a greenhouse. This modern material handling system cuts labor costs a lot."

• **Carts** are used to move Dutch trays. A cart can carry eight Dutch trays, 36 plug flats per tray or 278 flats of plugs at once. That's about 100,000 plugs. They are moved easily (on all cement walks) by one man. That's why it costs so little to move those plugs!

• **Cherry picking**. It's possible to remove one or several plug trays from the center of the bench or row of trays in the greenhouse. "It's a bit slow," Says Tom, "and is avoided when possible, but it can be done."

What are those different plug growing environments?

There are several—and nearly all are computer-controlled (Priva):

• **Temperature**. Both soil temperature (in the plug) and air temperature overhead can be maintained as desired in any house.

• **Light reduction**. Newly sprouted seedlings can't stand the strong April sun. So as light reaches a "set point" in footcandles, the computer draws 50 percent shadecloth over the crop, mostly on tender, newly sprouted plugs. These same shadecloth sheets also serve as energy curtains in winter. Again, it's all computer activated.

• **HID light** covers all plug beds—and they are again computer-controlled. Lights are on 18 hours in 24.

• **CO_2** is maintained at set levels (except during daytime ventilation)—again by computer. It's another of the many small things that taken together add up to quality plugs.

• **Many, but not all, trays** are first sprouted in three large germination chambers. Mist is provided to maintain saturated humidity in the chambers. It's also computer-controlled.

• **Humidity in the warm, moist** greenhouse is maintained by fog—especially over the first house as plugs leave the germination chamber. In later stages, watering is provided by ITS overhead boom irrigation. Again, it's sophisticated, capable of doing 20 or 30 feet, then skipping a piece, etc. The time and amount of water applied to the plugs is controllable.

So you see, this whole range of environmental "helpers," each vitally important to providing good plugs, is here at Wagner's. It's in place and working. Computer control eliminates a lot of harassing chores for management and workers.

Germination chambers

Smaller seeds are sprouted in germination chambers. Marigolds, tomatoes, etc. are germinated in the open greenhouse. After the plug sheets are seeded and watered, they are loaded onto Dutch trays, again, one car carries 278 flats. The cart is then pushed (next door) into one of the germination chambers. It may need only several days to sprout, and of course they must be watched several times daily to be sure they go out as soon as a satisfactory percentage has sprouted. They will stretch ruinously if left in a few hours too long.

Again, temperature here is maintained at optimum levels for each species. Relative humidity is near 100 percent. Light is provided for special things that respond to it, and mechanical refrigeration is available for cool-loving plants in summer (great for such things as pansies in July heat).

Putting it all together

From sow to sell here's how it all happens at Wagner's.

• **Seed sowing room** gets its orders each week from the control computer. Filled plug flats are delivered as needed, seeding is done (a Hamilton, a Vandana and a Blackmore), seeded plug flats are watered, loaded to Dutch trays, and off they go.

• **Chambers or greenhouse?** Trays now go either to germination chamber, or the easier things go direct to the warm moist greenhouse environment. For flats in the germination chamber, they are moved out promptly as soon as satisfactory germination has occurred.

• **Warm/moist/fog greenhouse environment**—newly sprouted seedlings go here first to get used to the outdoor world. The tomatoes, marigolds and easier types go direct from seeder to this warm, moist environment. As soon as enough root and top have developed to stand normal greenhouse conditions, they are moved to a second environment: less moisture, less heat. These temperatures depend, of course, a lot on the species involved. For example, in January there will be houses of begonias that need lots of moisture and heat. In fact, the size of this warm/moist environment area is itself flexible—can be expanded if needed.

• **A less moisture, less heat environment**. Next, trays move to an intermediate climate. Now ITS boom irrigation takes over—applying moisture/fertilizer as needed (judgment). Light is full sun for most species.

• **A normal 60-degree F environment** is the third step. This is a toning or hardening environment. Again ITS boom irrigation and plants get full sun.

Who decides which plug trays go where? Mostly Ron and his production people go through the operation daily "I just put a slip of paper on each lot so that our material handling people know where to move them—which environment," says Ron.

So there it is! The goal: to put each species in the optimum environment for that week. "We move the plant to the environment," says Ron.

Role of the computer

Pivotal!

• **Order entry**. As the orders come in, the order entry people feed them into the system. They have the availability of everything at their fingertips, along with weeks sow to sell for all species grown (much of the sowings are done to order).

• **Space needs**. The computer regularly provides total square feet of each environment needed week by week—to fill all customer orders on hand and to prevent overselling.

• **Instructions to crews**. The computer tells the seed sowing people how many flats of what to sow each week. It's all done on weeks numbered one through 52—mum style.

• **Shipping**. What to ship each week to whom. The computer supplies shipping documents, labels and creates invoices. And of course the office staff get data for accounts receivable records per individual customer.

• **Other crops**. In fact, all spring bedding plants, poinsettias and other flowering pot crops are computer planned in much the same way. Ron spends maybe half his time at his own terminal. There are six other terminals, so that everyone involved can work in the operation together.

Other points

Feeding is started as soon as seeds sprout—on everything.

Eight hundred per flat plug trays are used some. These tiny plugs mean low cost—but must be planted at once by the grower receiving the plugs. They're promising, but still not widely used.

Will the specialists win? When you see this intensive level of attention to detail, high level of automation, etc. possible with a qualified specialist, it makes you wonder whether plugs won't finally be done by specialists—as have mum cuttings and many other crops.

"

Vic Ball is editor in chief of GrowerTalks *magazine.*

High-tech plug growing succeeds at Four Star Greenhouse

December 1989

by Russell Miller

VPD irrigation and cool days/warm nights eliminates growth regulators, reduces runoff and provides for a higher quality plug crop. Take a look at hi-tech plug production.

Plug production is state-of-the-art at Four Star Greenhouse in Carleton, Michigan. The techniques used here to produce more than 27 million plugs a year put this company at the cutting edge of commercial plug production. Four Star Greenhouse is likely the only commercial plug grower in North America successfully using cool day/warm night technology and vapor pressure deficit irrigation on plugs.

"You have to stay out in front while maintaining quality and quantity," says Tom Smith, owner. "At Four Star Greenhouse, we keep on trying something new. You can grow quality plugs cheaper without this new equipment and technology, but if you plan to produce a consistent, high-quality, saleable plug crop year after year, you need to have the right tools, the right people and the right environment."

Germination chamber speeds plug production

The environment within the chamber is monitored and automatically adjusted by an Oglevee computer. The chamber is equipped with four misting nozzles, two on opposite walls, and an air over water fog generator. Because this germination chamber has insulated walls a half-foot thick, the computer can maintain relative humidity inside the chamber within 1 degree of where condensation develops.

The chamber has two cooling stages, using either outside air or inside air conditioning. An efficient air exchanger inside prevents air temperature stacking—meaning that no matter how many plug sheets are in the chamber, the air temperature is virtually identical at all vertical points from top to bottom. They can fit up to 6,000 plug sheets in the chamber.

Gary Vollmer, head grower, says the chamber is designed to maintain a perfect-as-possible environment, eliminate algae growth and reduce disease contamination to near-zero.

One of the first and foremost problems addressed with this chamber was condensation. Gary says the answer lies with the amount of insulation and the ability to maintain temperature and humidity to prevent condensation. The chamber's insulation has an R value better than 30, much higher than the average germination chamber.

Gary says they have seen significant improvements in the germination of many plug crops.

"We compared begonias, for example, germinated and grown on the bench to begonias germinated in the chamber and grown on the bench. The begonias germinated in the chamber reached a finished plug stage three days earlier than the begonias germinated on the bench."

A Blackmore seeder is used to seed the plug trays, using Greenway 12 percent perlite as the media. With the chamber, they save about five growing days of bench space for every plug crop.

"This allows you to grow more plugs using the same amount of space you already have." Tom says. "It buys you a little time."

State-of-the-art plug production at Four Star Greenhouse

Four Star Greenhouse is a third generation business that has been producing plugs since 1980. Spring production is half plugs, half finished bedding plants. In 1989, Four Star produced and sold more than 27.6 million plugs, ranging in size from 800-cell annual plugs to 50-cell geranium plugs.

Four Star also produced more than 70,000 bedding plant flats, 250,000 4-inch pots and 30,000 10-inch hanging basket geraniums. During the plug growing off-season, the company produced about 31,000 pots of Annette Hegg poinsettias for holiday sales. Crops are wholesaled in Michigan, as well as nationwide.

Four Star has two plug ranges totalling 44,000 square feet of floor area with about 33,500 square feet of bench area. Both plug ranges, as well as a new range being built, are Rough Brothers TechLite houses covered with 16- to 32- millimeter acrylic panels.

The plug ranges have three bays with concrete floors. One bay is 168 feet long and 17 1/2 feet wide, and the other bays are 180 feet long and 20 feet wide. The shorter bay allows condensation to drip onto the concrete floor, not the crops.

Four Star uses Rough Brothers RoFlo rolling benches with fin tube heating. Automatic irrigation is accomplished using three Grower Jr. booms from Integrated Tech Systems and three Andpros booms.

The growing environment is controlled with an Oglevee environmental computer that automatically monitors and adjusts air temperature, media surface temperature, irrigation water temperature and relative humidity. It's linked with an outdoor weather station and monitors light levels and evaporation rates within the greenhouses. It also provides on-screen graphical tracking needed for growing crops using cool days/warm nights.

Soil temperature in the plug growing ranges, under high light conditions, is kept 35 degrees higher than air temperature—both temperatures are monitored and automatically adjusted by the computer. The automatic boom irrigators apply water consistently and save a lot of time on labor.

A one-of-its-kind 20- by 40-foot germination chamber recently completed its first year of service at Four Star Greenhouse. Built within the seedling area, it features 6 inches of insulation—insulation boards with sprayed-in urethane foam. Water-proof fiberglass on the inside and outside walls make it easy to keep clean and disinfect.

Plug growing without growth regulators or stress

"Two years ago we got behind on our poinsettias, they were late and small," Tom says. "We expected to lose the crop. Then William Carlson and Royal Heins from Michigan State University came in and said we could save the entire crop with graphical tracking. We gave tracking a try and ended up with one of our best poinsettia crops, right on time. We are definitely sold on the DIF principle."

They used cool days/warm nights on 10 percent of their impatiens plug crop in 1988 and 100 percent of all their plugs in 1989. The technique works regardless of the different plug varieties being grown in the same greenhouse. They didn't use any growth regulators on the entire 1989 cool day/warm night plug crop—except on seed geraniums, where growth regulators are needed to open up the leaves.

"You don't have to stress the plugs by growing them dry in order to hold them back, either," Gary adds. "We can also grow plugs at a higher fertilizer rate, which promotes vigorous growth and more branching."

Gary says they had excellent results by combining cool days with natural high light afforded by the acrylic covering on the plug ranges. Tom and Gary say plugs grown in 1989 under cool day/warm night (negative DIF) are superior to plugs grown the year before under conventional warm days/cool nights (positive DIF).

"The drawback is that the plugs grown under negative DIF have a different appearance," Gary says. "They have a light green color—there are no deep green leaves. The leaves also droop downward and then tip up at the ends. Once you remove these plugs from negative DIF they quickly come out of it and the leaves turn a deep green. They grow more vigorously than a plug grown under conventional temperatures, and the finished plants are more compact with more branching than a plug grown under normal temperatures using growth regulators."

At night, the greenhouses are heated, if necessary, to create the warm night effect. Vents are opened at sunrise, bringing the inside temperature down several degrees or more than the night temperature. The costs saved by using no growth regulators is slightly offset because Four Star Greenhouse also incurs higher heating costs.

"If you would have asked me two years ago whether I thought we would be using cool days/warm nights on poinsettias and plugs this year, I would have laughed," Tom says. "But we are sold on it. We are really pleased."

VPD irrigation—the most exciting concept for plug growers

In 1988, Four Star Greenhouse experimented with vapor pressure deficit (VPD) irrigation on one bay of plugs. It's so successful that they are now using VPD irrigation on all bench crops and plugs in the germination chamber. Next year, they hope to be using VPD irrigation on all hanging baskets. VPD is used two ways: On the bench crops it's used to accumulate the values needed to determine when to irrigate the crops. In the germination chamber it's used to control humidity.

"VPD irrigation is the most exciting thing to happen in the industry," Tom says. "I firmly believe it is the future of commercial crop production. It is the most accurate way of monitoring irrigation water. We feel real confident using VPD irrigation with the computer, especially in the germination chamber during propagation. It's excellent. With VPD irrigation, you never have to worry about overwatering or underwatering a crop. It's an exact science."

"VPD is definitely the trend of the future," Gary says. "VPD irrigation has high potential in propagation. The computer automatically and precisely adjusts misting according to the environment. With the computer doing the work, it's hands-off control. There is less risk of root rot and bacterial problems. You don't have to adjust misting to the crop—the computer does.

In order to incorporate VPD irrigation on their bench crops, the staff first sets a target amount for watering each crop. For instance, it was determined that each 6-inch Annette Hegg poinsettia in the early growth stages at Four Star requires 9 ounces of water per watering. This was accomplished by handwatering the crop once to determine the correct amount of water to be applied. The amount of water needed changes according to the crop, the growth stage of the crop, container size and environment.

"Once you set a target for the computer to operate on, it automatically adjusts irrigation to meet the requirements of the crops until finishing," Gary says. "We're just beginning to tap into VPD irrigation. On the difficult-to-germinate plugs, we are finding that VPD really pays off. We can achieve noticeably high germination rates on the difficult-to-germinate material."

New range to be unique
A new one-half-acre Rough Brothers TechLite range being built at Four Star Greenhouse will be unique in that it will have flooded floor irrigation. During the height of the growing season next year, bedding plants will be grown on the floor while plugs will be grown on the benches. Poinsettias will also be grown on the benches and the floor.

"What we will try to achieve with this new range are primarily two things: We want a range that allows for flexibility and a range that will not produce runoff," Tom says. "The irrigation water will be recirculated. We want to address the environmental problems associated with greenhouses long before the EPA comes in and addresses the problem for us.

"You can change your growing methods to fit your greenhouse environment," he adds, "or you can shape your growing environment to match what the crops actually need—which is what we do at Four Star Greenhouse."

"

Russell Miller is features editor of GrowerTalks *magazine.*

Under an acre

December 1989

by Vic Ball

Profile of Ivey's, Lubbock, Texas

The Iveys are a well-done 40,000-square-foot retail growing range in Lubbock, Texas. Glen Ivey Sr., son Mark and grower Chris Spray do it.

Most interesting is their plug production. They grow most of what is needed for their 30,000 flats (do buy in some). It's a budget set-up, using off the shelf heat, mist, etc. It does not give 95 percent plug sheets, but says Glen, "By overplanting a bit we get all the plugs we need." Important: The initial investment is modest. And watering is largely automated. How's it done?

Benches and heat
The bench is a typical cement block two by four structure with treated planking wood floor and sides. The bench planks are spaced to allow heat to come up from below and for drainage.

Heat is supplied by a fan jet-type burner/blower set on the ground at the far end of each bench. Heat blows under the bench through typical 30-inch poly tube with holes. The plugs are all grown on two and a half benches, each 10 by 120 feet.

The heat is designed so that different temperatures can be maintained in each bench. The burners/blowers burn propane. Skirts are hung from the side boards to keep heat in below the bench. It works. In fact the same type of underbench heat is used throughout the whole 40,000 square feet.

Misting is simple and practical
Next, the misting systems. Again, it's simple, practical, effective. The Iveys

have installed nozzles on about an 18-inch riser above the bench with a nozzle at the top of each. They are spaced about 3 by 5 feet allowing ample overlap and no dry edges. Two rows of nozzles do a 10-foot bench. The nozzle they use is a Pate nozzle K-15 from Sharp in Seattle, Washington. Water pressure is maintained at 30 to 50 pounds. The installation is all piped with PVC.

There is a mist control panel that controls the irrigation. With this they can mist newly sown plugs on a warm day up to two seconds every four minutes. Misting can be gradually scaled back to a few seconds every two hours.

Chris says, "You've got to watch those sheets carefully. The sudden full sun can dry out newly sprouted seedlings fast. I check the mist control every morning and every noon. We do very little handwatering, but you must keep an eye on them." Again, it works.

Most plug sheets are 390 per 21-inch flat. They use Ball Germinating Mix for plugs. Fertilizer is included with the mist about 10 days after sowing at half strength. Greenhouse temperature is normally maintained about 68 F at night. For soil in the plugs, their target is about 70 F.

Plugs cut time in half

Says Glen:"Since we started with plugs, we've cut our transplanting labor in half. We set the filled flats out on the bench, the crews transplant them right there." Cost for all of this—modest. Here are some rough figures:

• **Cost for heating one bench:** About $510 for the heating burner/blower, $25 for the 18-inch poly tube.

• **Misting equipment:** Nozzles for a bench were about $700. The mist control system was roughly $250. Plus PVC pipe, etc.

Second retail outlet sells more

Besides their 1-acre retail growing range, Iveys operate a 6,000-square-foot retail outlet (glass roof) at a local shopping center.

On containers: The standard fare at most Texas bedding plant ranges are the little 1½-inch pots. You set 50 individually into a flat and fill them. They have been used for years and are still standard on many ranges. Iveys have broken with all this and use cell packs, 72 per 21-inch flat.

Speaking of flats, Glen reports that they sell a lot of full flats ($17.90 retail). Sales of full flats go up every year."Demand for everything last spring was quite good. We pretty well sold out. About 60 percent of our pack annuals are sold as full flats. We do allow mix match. We do a lot of business with landscape contractors. We are not excited about an automatic plug transplanter—we can get plenty of good help here."

Trends at Ivey's

"We have already doubled our operation, we're about 72,000 square feet today and have also doubled our plug production area," Glen says. "By the way, we've gradually gone from all wholesale in 1986 to mostly retail this past spring." More profitable!

"Perennials are just not a big seller here. Our big three bedding plant items would be vinca, petunias and portulaca." Glen says. "All of them will stand our terrific heat down here. Marigolds are also strong, and jalapeno peppers are very big. This is cotton country—not much unemployment but we seem to get the help we need. People are prosperous here, which helps us."

"

Vic Ball is editor in chief of GrowerTalks *magazine.*

Subirrigating plugs: Is it the future?

September 1989

by Debbie Hamrick

Subirrigation is becoming standard practice for pot crops and even some bedding plants. Dutch growers especially are moving in masses toward subirrigating with ebb and flow on concrete floors. But subirrigate plugs, and on the floor?

That's exactly what two large Dutch plug producers, Hamer in Zwijndrecht and Zaadunie in De Lier are doing. They're adopting ebb and flow floor system technology to plug crops and it's working. Other growers are adopting this technique for finishing bedding plants. John Ammerlaan of Leo Ammerlaan b.v., Berkel, is one Dutch grower using subirrigation to irrigate bedding during finishing. Here are highlights from visits to these three production facilities in May.

Background details

The standard Dutch ebb and flow system for the floor is sectioned into parcels 6.4 meters wide (about 19 feet). This is the distance from one gutter to the next in a standard Dutch house. Length of the watering section varies with length of the overall width of the greenhouse. Sections extend perpendicularly from center aisles. The goal is to keep watering sections small so that filling and draining is as quickly as possible to get the job done. Water stands in the basin for 10 to 15 minutes until all containers are watered. Each section is surrounded by a raised concrete lip; this forms a shallow basin. Water enters and drains from the center of the basin, through a lengthwise gutter. The floor is sloped slightly from outside, lateral edges toward the center gutter. Precisely grading the concrete is vitally important to avoid puddling.

There's no question that subirrigating with ebb and flow saves labor, but there are drawbacks and concerns to such a system, as well.

Points in favor of subirrigating on the floor

Flexibility in container sizes and spacing with 100 percent space utilization is the biggest advantage of using ebb and flow on the floor.

Herman Hamer, Hamer's, says that ebb and flow saves "a lot of labor on watering." Having no aisles is not really a problem, he says, because plug trays are pulled middle row first onto CC carts. "You take the cart with you so you don't have to walk back and forth as you would with a table system, and we are able to use almost all the space we have."

The system on the floor is also more economical than installing an ebb and low system on moving tables, he says.

At Hamer's many crops, including impatiens, begonias and petunias, start off in a germination room with fog and fluorescent lights before they go to the greenhouse. Plug crops are also under supplemental lights in the greenhouse—7,500 square meters (80,600 square feet) of supplemental lights in all. Also, in the beginning, plugs are handwatered, not subirrigated. Media for plugs is compressed peat with perlite added for aeration. "That makes it easy for filling gaps," he says. Hamer's is also working to develop a mechanical system that patches empty cells, which requires packed plug cells.

At Zaadunie, flexibility is also the reason they selected an ebb and flow system for the floor to grow plugs in their new 10,000-square-meter (107,500

One of the goals of an ebb and flow system on the concrete floor is to keep the area to be watered small. At Hamer's each basin is 250 square meters (2,600 square feet). The floor is sloped—$1/2$ centimeter from outside lateral edges toward the center. Heat pipes in the floor, used only for the first weeks of plug growth, are spaced 10 to 15 centimeters apart (4 to 6 inches). Hamer's uses two other above-crop heat systems during growing on and finishing. One of the heat systems overhead also doubles as support for a monorail transportation system. The other system, a fin-tube system shown in the photo, is suspended from cables. Each evening, the grower lowers fins to about 1 meter (39 inches) above the concrete. During the day, the fins are stored overhead. Herman prefers the fin system for the excellent air circulation it provides and uniformity in drying out trays.

square feet) production house. Their research division has been testing ebb and flow for plugs for more than a year.

"Subirrigation with rolling benches and the whole logistics of that doesn't work for us," Martin Uittenbroek, manager, says. Logistically Zaadunie deals with some 5,000 different sowing dates, up to 800 individual production items, and during peak shipping may ship up to 50,000 trays in a week. Being tied to benches just didn't add up.

The new structure will see its first trial crop in December. "We learned from others that we may have a problem with pH, so we want to run a trial first." They will grow peppers and tomatoes initially, before working with floricultural crops on the new system.

Other factors in favor of subirrigating on the floor include:
● **Ease in recirculating water** without all of the separate pipes and fittings required for ebb and flow table systems. Because floors are concrete, you control where the water is going.

Martin Uittenbroek at Zaadunie expresses concern, however, over pesticide residuals in concrete. Bonzi, he says, seems to be residual in the concrete. Since Bonzi is very active at low concentrations, this could cause major problems, especially with plugs.
● **Sanitation during production** and after production is over is easy. Foliar disease control is also easier: Since crops are watered from below, foliage is always dry.

Why not subirrigate on the floor?
● **The biggest drawback** to subirrigating on the floor is that workers must stoop to place pots, space or to pull orders. For spacing-critical pot crops, this is a major drawback. For long crops with only one spacing or crops that don't require any spacing, it's bearable.
● **Heating**. Most of these systems are designed with in-floor heat systems. This is good and bad. It's good because it encourages root growth when crops are first being established. Heat in the floor also helps in drying floors quickly. But, in-floor heating systems take a long time to raise or lower the temperature to set points, and heat buildup especially can be a problem.

Herman Hamer prefers to use the floor heat system for the first three weeks of plug crops only. "Using heat from the floor only makes soft plants with poor roots," he says. Also, Herman says heat buildup under plug trays is very difficult to control. Hamer's has two other heat systems in addition to the floor system. They have overhead metal heat pipes that run the length of bays that also support a monorail transportation system. The second system they use is an over-the-crop aluminum-fin system that raises and lowers on cables. The fin system is stored overhead during the day, but lowered in the evening to about 1 meter (39 inches) above the concrete. Any lower, he says, and trays would all dry unevenly. The fin system allows for optimal air circulation.

John Ammerlaan, who subirrigates bedding plants during finishing, uses his floor heat system for the first week after plugs are transplanted into the growing container only. When plants are rooted, he switches to overhead heat.

● **Media in a subirrigation system** is critical. Jacob ten Wolde, head of production development for Zaadunie De Lier, says the ideal media for subirrigating plugs should totally dry out in 48 hours, "so you can water again. You want a coarse media with good capillary action, but not one that dries out too quickly." If media dries too quickly, it isn't good for shipping.

Zaadunie's De Lier range has been experimenting with ebb and flow subirrigation on the floor for plugs in a 1,500-square-meter (16,000 square feet) research greenhouse. They've designed a special "shuttle" that plug flats fit into. Dimensions are 47 by 37 centimeters (18.5 by 14.5 inches).

Shown in the inset upsidedown, the bottom of the shuttle tray is channeled so that irrigation water may easily pass through. Since there's a lip as well to keep the bottom of the plug trays up and off the floor, roots are airpruned. Shuttle trays stack—for going into germination chambers or shipping. The design allows for excellent air circulation when stacked. Plug trays, which nest inside, are 538s (multiseeded crops), 219s, 106s or 78s (cyclamen). They also are experimenting with a 48-cell tray for cyclamen.

Jacob has reservations about perlite for increasing aeration. And, based on his tests, he's found that Oasis has not worked well for subirrigating plugs, and rockwool is too expensive to be practical. Also, when rockwool is transplanted into media in the final growing container, it dries out too quickly.

Because plug trays must be economically patched to fill empties, compressed peat cells is the media of choice at Zaadunie. "For subirrigation you don't want dust in the peat. Good quality peat is important." Before seeding, each block is punched with a hole so seed is placed in the center of the cell.

"Pressed pots are settled in the marketplace," Martin explains. "In theory there are disadvantages in pressed pots for subirrigation, but in practice the disadvantages are not there so much. In begonias you can see a negative difference, but not for other crops."

Jacob adds, "In tests we've seen that pressed pots are just as good as loose filled cells. Pressed pots give more uniform watering. But, when pressed pots totally dry out, they are very difficult to water again."

Overall, Jacob's research has shown that subirrigating plugs speeds crop time—even in finishing for the grower in some cases. It provides uniformity in watering and fertilizing. "We believe subirrigation is the future," Martin says.

"

Debbie Hamrick is editor of GrowerTalks *magazine.*

Florida bedding plant and plug producers are geared for change

June 1989

by Russell Miller

Florida update: Expansion and renovation, Federal Express, water restrictions, 4-inch and 6-inch bedding, more and more plugs . . . these are changes occurring in south and central Florida at three wholesale operations.

It's never business as usual in Florida. Year after year, Florida bedding plant growers find something better, something less expensive, something more efficient. From the perspective of three wholesale operations growing plugs or finished bedding plants in central and south Florida—Natural Beauty in Apopka, H.M. Buckley & Sons Inc. in Naples and Knox Nursery Inc. in Orlando—come these updates.

Natural Beauty starts summer plug program

Natural Beauty of Florida in Apopka, with 22 acres under glass, is one of the most modern ranges in America. The majority of the plugs, marketed as Sparkplugs, and the larger-sized Expeditors are sold through Ball Seed Co. and are boxed and shipped directly to growers nationwide. Through Ball Seed, Natural Beauty has initiated a summer Expeditor plug line consisting of heat-tolerant varieties.

Natural Beauty will produce more than 200 million Sparkplugs and Expeditor plugs this year. This is an increase over last year's production, with future increases being predicted.

Ron Derrig coordinates the summer Expeditor program, which includes the Pacific abelmoschuses, Flaming Fountain amaranthus, gomphrena, *Zinnia linearis*, Sunnyside portulacas, and Carpet and Little vincas.

"These are just a few of the outstanding varieties that are proven to be heat-tolerant for summer production," Ron says. "These plants are solar-charged under the long, intense light days in Florida and are acclimated to the hottest regions in the United States. This product line provides bright, vibrant colors during the dog days of summer.

"We have seen an increased demand for both Sparkplugs and Expeditors over the last few years. Growers have experienced the difficulties of growing their own plugs and are now turning to the plug specialists. They have also realized the cost involved in growing good, quality plugs, and the profit they have lost by displacing a finished bedding plant flat with a flat of plugs. Many growers are buying plugs for their late season crops so that plug trays are not taking up valuable space during the peak spring season."

Growers are looking for more knowledge concerning transportation and product handling, he adds. They want to know more about how the plant material was grown before it was shipped, what fungicides, insecticides and growth regulators were used, and when they were applied.

"This information," Ron says, "is becoming more available. Research is now geared toward post-shipment of plugs. This will help the customer get the plugs off to an even better start and create a higher quality finished product."

Knox Nursery expands GoPlug facilities and production

Knox Nursery Inc. in Orlando is increasing its GoPlug production facilities from 85,000 square feet to 115,000 square feet this spring. "We never anticipated such high demand for plugs from other growers when we produced our first plugs in 1986," Bruce Knox says. "Since then our accounts have increased considerably." Vaughan's Seed is the company's No. 1 account.

Bruce's father, Jim, says they plan more design changes in the coming years to meet the rising demand for GoPlugs. "We are not finished with our plug production facility. We originally designed the facility for our own plug production, not expecting to end up growing plugs for other growers."

Today they are producing 26 bedding plant varieties, plus vegetables, in 392-plug trays, sold as 375. Geranium plugs are grown in a 392-tray sold as 350.

Four-inch is a "hot item," Bruce says. "We are seeing more 4-inch than cell-packs everywhere these days, even in supermarkets.

One of the most modern plug production ranges in America, Natural Beauty's bench system on rails facilitates moving plug benches from seed to boxing and shipping. Plans are underway to further increase the system's efficiency.

Federal Express is moving more plugs

Ron Derrig says about 65 percent of the plug orders received at Natural Beauty are shipped Federal Express. The rest is shipped via truck-to-door delivery, with a small amount of customer pickups at Apopka.

Advantages of using Federal Express at Natural Beauty are two-fold: Not only does Federal Express guarantee delivery direct to the customer within 48 hours, but plant material grown at two other operations operated by Greiling

Farms (in Denmark, Wisconsin and Blairsville, Georgia) can be shipped from any of the three locations on the same day, and delivered to a customer all at the same time.

At Knox Nursery, although pleased with the Federal Express shipping system they are using, Bruce says there needs to be alternatives to shipping through Federal Express because the costs are high. "We pay a premium price to ship Federal Express. But today, there are no other alternatives available." About 80 percent of Knox Nursery's 30 million to 40 million GoPlugs produced each year are shipped through Federal Express.

Southwest Florida's water shortage is critical

The region surrounding Naples, Florida is experiencing what's called a 100-to 200-year drought, meaning that the water shortage is the worst that's occured here in the last 100 to 200 years.

"It's a sign of what's to come in many parts of the nation," says Jim Pugh, general manager of H.M. Buckley & Sons Inc., a 23-acre wholesale bedding plant operation with nearly 15 acres in production. A television news team visited H.M. Buckleys during mid-April as part of their coverage of the south Florida water shortage crises.

Jim has been meeting with officials from the local water management district during mid-April trying to convince them that nurseries such as H.M. Buckley & Sons are using water for agricultural purposes, not urban and recreational purposes—the category H.M. Buckley & Sons is presently listed in.

Under the present Phase 2 water restrictions in the Naples area, water use restrictions are far more restrictive for urban and recreational purposes than they are for agricultural purposes. "The major problem with this categorization is that we aren't allowed to water at our discretion," Jim says.

Plugs being boxed at Natural Beauty come via an efficient rolling bench and conveyor system. From this point, plug boxes for Federal Express shipping are stacked near one of several docks, ready to be loaded into semi trailers.

The water district isn't taking into consideration that different businesses have different water needs and they are grouping several types of businesses in the same category.

H.M. Buckley & Sons has also had to reduce the amount of water pumped from their own wells by 30 percent, and businesses under the urban and recreational categorization are also only allowed to water for no more than 12 hours every other night with an unrestricted amount of hand watering during the day. The water restrictions are enforced by local law officials and violators can be fined and imprisoned.

With 15 acres in production, handwatering is possible at the nursery, but requires too much labor.

Because of H.M. Buckley & Sons' past water conservation efforts at the nursery, the water management district gave the company a variance by allowing watering for six hours every night.

PowerShipping through Federal Express has its advantages

Natural Beauty of Florida has three onsite Federal Express stations, and both Greiling Farms in Wisconsin and in Georgia have one each for a total of five stations. Greiling Farms Inc. worked very closely with Federal Express Corp. in developing the shipping systems now used by other Florida plug producers.

Knox Nursery has had their onsite Federal Express station since August of 1988. Speedling Inc. in Sun City also had one installed last summer. A plug grower doesn't get an onsite Federal Express station simply by requesting one. Federal Express Corp. determines which businesses should have an onsite station in accordance to what will be most cost-effective and efficient in the Federal Express worlwide shipping network. In most instances, you have to be shipping at least 20 packages a day through Federal Express before they consider installing an onsite station.

From an on-site station, large shipments are loaded onto semis headed toward Memphis, Tennessee. Small loads are moved by vans to the nearest Federal Express terminal, most likely an airport at Orlando or Tampa, for example. From there, plug cases are either loaded onto semis or are transported by air to Memphis.

Federal Express installs, free of charge, the PowerShip computer system at onsite stations. The PowerShip system is a "supermeter," enabling users to generate shipping labels and management reports. When an order is received from a grower, the information is logged into the PowerShip system. The PowerShip system also generates the shipping label for each plug case, along with an invoice.

The invoice generated is an invoice from Federal Express to the plug producer showing shipping totals (accounts payable). It's not capable of generating customer invoices (accounts receivable).

The shipping label on each plug case has a bar code containing all necessary information in order to deliver the package properly. Plug cases are stacked together at a central area at each nursery for pickup by Federal Express drivers. Federal Express loads the trucks and, once out of the nursery, Federal Express never uses conveyors to move most plant material, including plugs. This eliminates chances for package damage which can occur on conveyor systems.

Federal Express drivers have a handheld scanner that is used to scan the bar codes on every plug case prior to each case being hand loaded into the Federal Express vehicle. This scanner is then plugged into an on-board computer located in the cabs of Federal Express vehicles. The information is transferred to this on-board computer and into the Federal Express worldwide computer network.

The Powership computer system is part of Federal Express' customer oriented service and management operating system (COSMOS), based in Memphis. Information from every Power-Ship system is transferred nightly into the COSMOS system. Through COSMOS, Federal Express is able to track any package instantly simply by inputting the shipping label number.

With the onsite PowerShip stations, a plug producer has the ability to trace any Federal Express shipment instantly. One advantage is that the producer can locate any order during shipment and have automatic proof of delivery to the customer. Every time a plug case is handled by Federal Express personnel using the handheld scanners, the shipping information is transferred into the worldwide tracing network.

Federal Express guarantees, for any person shipping Federal Express, that they will locate and tell a sender exactly where any package is during shipment within 30 minutes of inquiry. If they can't, the full value of the shipment is refunded.

For plug producers, plug cases are sent Standard Air Service, which delivers the plugs to the recipient within 48 hours after the plugs leave the nursery. Federal Express guarantees delivery by 5 p.m. the second day, and if it doesn't reach the correct destination by 5 p.m., even if it's a minute late, Federal Express refunds the shipping charge. Every plug case sent by Federal Express is valued at $100.

The advantages of shipping through Federal Express stretch to every grower who receives a plug case from Natural Beauty, Knox Nursery or any other plug producer using Federal Express, which is able to deliver next-day service to more than 99 percent of the nation. The grower gets a fresher plug, and the plugs are delivered right to the door within 48 hours after they leave the nursery. The plugs also arrive undamaged and on time—it's guaranteed.

Refunds are very rare and claims are of a relatively small percentage, Bruce and Jim Knox say. "We ship several hundred plug cases a day during the early spring and the system works great," Bruce says. "To this point, we have never had any problems."

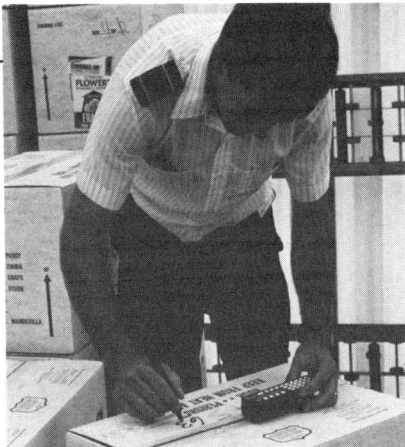

"At this time (late April), H.M. Buckley & Sons is finding it virtually impossible to adhere to our current variance under the Phase 2 restrictions," Jim says. "We must have the flexibility to overhead irrigate critical crop areas during the day at our discretion. If we had this flexibility, we are certain that it would result in a significant overall reduction in our water consumption."

The company has 2 acres of poly covered greenhouses and 6 acres of shadehouses. With 7 acres of outdoor production under the Florida sun and a narrow window in which to sell finished bedding plants at this time of year—seven to 12 days under normal conditions—Jim's work with the water management district has been crucial in minimizing losses.

"We have self-imposed plant quality standards at H.M. Buckley & Sons that we are going to keep," Jim says. "We just sustained a $100,000 loss because we couldn't adequately water our bedding plants and provide the quality we want to produce."

6-inch summer crops are growing in demand

This year H. M. Buckley & Sons are growing and shipping roughly 4 million 4-inch, 18-pack trays and about 650,000 6-inch pots to out-of-state Kmarts and to all Builder's Squares in Florida. Both chains account for 70 percent of spring production and the remaining bedding plants sell to Florida landscapers and developers.

Kmart and Builder's Square are basically the same company, but they operate independently of each other. This spring Kmart requests a UPC tag in every pot; Builder's Square wants a SKU tag in every pot. So H.M. Buckley satisfied both with a UPC tag and a SKU tag in every pot, with different code numbers. Kmart also wants to use as few bar code numbers as necessary, so all 4-inch crops sell at the same retail price. All 6-inch, regardless of variety, are also priced the same.

"Two years ago we didn't grow 6-inch bedding plants," Jim says. "Our 6-inch program for summer-type annuals is expanding and there's a drop in 4-inch on some varieties. We still do a large amount of 4-inch in what we call a 4-inch fill, a nice chain store item, but we try not to duplicate them in 6-inch.

"This is because we are seeing a trend in some areas where the same variety in a 4-inch competes against the same variety in a 6-inch when they are on the same shelf. So we are zeroing in on varieties most suitable for 6-inch production, not marigolds or salvia, for example, and not producing them as a 4-inch to avoid the competition between the same varieties."

Impatiens now account for 35 percent of spring production and almost half of fall production. There's a continued rising demand for dwarfed hibiscus, Jim says. H.M. Buckley & Sons is growing 250,000 dwarfed hibiscus as 1-cutting per 6-inch azalea pot, pinched once with applications of Cycocel.

Loading trucks within an hour and a half

Independent truckers arranged through a brokerage firm and H.M. Buckley & Sons' own trucks deliver the company's plants to about 380 destinations throughout the East and in Florida. Usually during the peak shipping season, March through May, three to four trucks for out-of-state deliveries and four to five trucks for local deliveries are loaded per day. Most out-of-state trucks have 46-foot or 48-foot temperature-controlled trailers.

H.M. Buckley trucks—five trailers and four straight rigs—have two-by-fours spaced 6 inches apart running lengthwise down each side of the trailer. Other trailers used by the company have A-tracks inside, onto which two-by-fours can be hooked.

With trailers having A-tracks, workers can lay sheets of plywood across the two-by-fours. For shelving in company straight rigs and trailers, they lay two-by-tens across the two-by-fours.

Boxed plants for out-of-state sales are stacked into the trailers; about a thousand boxes will fit into a 48-foot trailer depending on the box size—but some 6,700 18-packs shipped unboxed can be loaded in the same trailer.

"We try to have all out-of-state deliveries made within 24 hours after they leave the nursery and all in-state deliveries made within 24 hours after the order," Jim says.

99

Russell Miller is features editor of GrowerTalks *magazine.*

Automated plug growing for under an acre growers

October 1988

by Vic Ball

Half-acre to 1 acre growers *can* produce good plug sheets. The important news is that new irrigation equipment will do the misting and watering job for you—reliably. It works. "It waters and feeds much more evenly than I can do by hand," says one grower.

Ray Banko in New Jersey, who grows 75,000 flats, says, "I wish I had been doing this for 20 years. It works just fine and I don't have to be in the plug house all day with a hose." Here are details on two success stories:

Tindall Greenhouses, Hightstown, New Jersey

Andy Tindall and son Larry do 15 96-square foot houses of bedding in the Hightstown, New Jersey area, all wholesale. One of their 30-foot by 96-foot houses is the "plug factory." Typical of several growers we've seen, they are using the porus cement floor (a high-sand cement mix); details from Bill Roberts at Rutgers, New Brunswick. In this floor are buried $3/4$-inch heavy-duty black plastic pipes every 10 inches. The hot water temperature is thermostatically controlled so that the soil temperature can be held at very close to the desired 75-degree F temperature—temperature control is reliably automated. The plug sheets are grown right on the cement floor.

The other major labor saver here is the new boom irrigation systems. There are several. Tindall's is using Integrated Tech System's irrigator (call (609) 448-6533). It moves a single boom down the length of the 100-square foot house and back. Since there are a lot of nozzles (24 on a 30-foot boom), water distribution is very even.

"It's a lot more even than I can do by hand, really," says Andy. "And that goes for the fertilizer that goes on with the water. I'm really proud of my plug flats. With that many nozzles, there's so much overlap that the water has to be evenly applied. I set the boom to travel up and down the house as much as every 30 minutes in warm, sunny weather—less often on cooler, cloudy days. The equipment permits watering only parts of the house and skipping over other parts as needed. Also, if something goes wrong—if I leave a wheelbarrow in the way—it will turn the water off automatically.

"The cost to equip a house with this system today is about $3,300. It was less when I put it in several years ago. We produce about 80 percent of our plugs this way and buy in the rest. Our begonias are from pellets—which works fine if you sow the seed within several weeks after you get it."

"We also have the same boom irrigation system in each of our 16 bedding production houses. We've converted our place over from hand watering over a period of six or eight years—several houses each year. Especially with the eal problem we have of getting help in this part of the world, it's been a real blessing to have at least most of the watering of the bedding plant crop automated. We do some hose watering on the bedding plant crop that's automated—some hose touching up on the edges—but the great majority of the job is done by the boom irrigators in each house. The Integrated Tech System seems to work very well—there's no important problems with it."

Ray Banko, Dayton, New Jersey

Ray is a 3½-acre bedding plant grower who does about 75,000 flats a year, heavy on impatiens. For the past several years he's been producing his plugs with a system that's basically the same as Andy's system—and he's very. pleased with it.

Ray uses two 25-foot by 125-foot greenhouses, also porous cement floors with PVC pipes every 10 inches. He emphasizes the importance of space heaters "to keep control of moisture in the air," he says. "Humidity tends to run up in plastic houses during the warm spring days." He has no trouble keeping an even 75-degree F soil temperature in the plug trays throughout spring.

Irrigation is again one of the new boom systems. This time it's from Growing Systems Inc. (Call (414) 263-3131). The boom travels up and down the house as often as desired—as much as every 30 minutes on sunny days. Liquid fertilizer can be applied through the misting nozzles and most of all, it works just fine. Ray reports his plug sheet quality has been very good. He moves the plug trays out of his houses as soon as they are well through the ground. Then he puts them in a cooler house with less moisture.

Ray and other growers we've seen often install a third layer of polyethylene below the double poly roof to prevent drip from washing out areas of the plug sheets. This third layer is suspended 3 or 4 inches below the normal double roof and carries the drip down to the edge of the house.

Again, Ray produces the majority of his plugs and buys a few things to fill in each year. Other bedding crops are hand watered.

Carl Blasig, Hightstown, New Jersey

Carl has also installed a boom irrigation system for producing plugs and again, he uses buried heating lines. It's much the same system, except this system is under glass. Carl reports no problems in producing most of his own plugs for 1 to 1½ acres of bedding plant production.

Interesting point here: Nearly all of Carl's bedding plant production is boom irrigated. In this case it's the system that carries a boom up one house, moves it over to the next house, and so on through 1 to 1½ acres of production in about five hours. Again, Carl reports that the boom does a good job. It is the Integrated Tech System's irrigator which provides maximum flexibility for watering partial houses as desired and for skipping certain areas if you wish.

Also, if something gets in the way of the boom it will stop and shut the water off—and announce to anyone nearby that there's a problem. Carl has been moving steadily toward boom irrigation on spring pot crops such as 4-inch mums and even poinsettias in the earlier stages.

He is still hesitant about overhead watering pot mums as they come into color—the same for poinsettias. He uses spaghetti tubes on the poinsettias after the final spacing, but is using boom irrigation on poinsettias in the earlier stages, including the very critical rooting period.

Carl says: "Because of the eveness of moisture application with this boom, we find that rooting poinsettias actually works better with boom than with hand watering or fixed nozzles. The fixed nozzles tend to have an uneven pattern of moisture distribution that makes for problems. The boom lays down an even sheet of water. That's important for any crop."

Conclusion

Certainly there are many other growers doing their own adaption of this basic system. But it seems that with reliable soil temperature control and reliable boom irrigation, that under-an-acre growers can do plugs very nicely.

"

Vic Ball is editor in chief of GrowerTalks *magazine.*

Blueprint for plug success
A look at Raker's Acres

June 1988

by Debbie Hamrick

There are many reasons why Raker's Acres, Litchfield, Michigan, has become one of the most respected plug growers in the U.S. Probably the most important reasons are their staff's devotion to detail and a company commitment to adopt new techniques that take them closer to the ideal plug growing environment or improves handling efficiency.

Raker's Acres is a family affair—brothers Gerry (administration) and Dave (growing), with nephew Tim (shipping) and Gerry's wife, Pat. They used to be Detroit area truck farmers but moved to rural Litchfield as urbanization spread. Last year they still farmed 1,100 acres—corn and processing tomatoes—but the emphasis is on being plug specialists, they don't even finish bedding anymore.

As truck farmers, the Rakers turned to bedding to extend their production season and increase cashflow. They worked up to producing around 25,000 flats annually. When they began working with plugs for their own bedding, they realized their energies were being split into too many directions. Plugs got all the attention until transplanting started; then the emphasis shifted to finishing flats. Plug growing suffered, and to top it off, they were still farming. Something had to give. They decided to specialize and ride the wave of the trend toward bedding growers buying in plugs.

Plugs have changed the Rakers. they've opened up bedding plant production doors for small retail growers with only a few thousand feet, Gerry says. Cut flower growers producing seed crops also prefer plugs because of the uniformity of crop harvesting.

The plug growing range covers two acres, with another one acre of freestanding houses for vegetable transplants.

Volume of plug production is around 35 million (ornamentals only) yearly, shipped nationwide direct and through brokers. The average order is around 30 trays. Around 10 percent to 15 percent of their business is specialty sowing (minimum order of 10 trays). While some larger growers tend to buy in only those difficult-to-produce items, Gerry says smaller growers order everything in plugs.

To increase their own space efficiency year-round, they're beginning to work with cyclamen—supplying both plugs and prefinished to the warmer, southern market where it's difficult to grow a quality crop from seed, but cool fall nights are ideal for finishing. Another crop item they've added to their roster is vernalized perennials in 120-cell trays.

Computer ties departments together

In December, Raker's went to a multi-user, custom-programmed computer system that ties together receivables, payables, order entry, inventory and production planning. There's a total of five terminals in the operation—three in the office and two in the greenhouse, one for shipping and one for production. The shipping computer terminal is conveniently located near the Federal Express work station.

As plugs are ordered, the computer automatically enters space reservations for the booked crops through production. Since "every square foot is usually booked by November" as Gerry says, it's then just a matter of concentrating on production.

During the season, Gerry can call up a tally of exactly what Raker's is committed for, when it was or is to be sown and when it will be shipped. As trays go in or leave the greenhouse, the computer automatically updates the amount of available space.

For example, if Gerry wanted to ship 1,000 6-inch prefinished cyclamen on September 30, he enters the information as a booking. The computer then automatically backtracks through production and subtracts space for the order through all production phases, from the initial plug trays, through the 2-inch plug phase, then into the 6-inch rim-to-rim pot phase and later the 1-foot on center phase. For regular bedding plugs, the computer also compensates for time of year and variety—these differences however, are becoming less and less from summer to winter Gerry says. With tighter environmental control, plug finishing times are evening out through the seasons.

Next year, Raker's plans to incorporate barcoding with the computer system. Before each plug tray is filled it will receive a barcode with a sequential number. After filling, the tray will be scanned and tied in with a particular seed lot. The code will allow the computer to keep track of each tray, and its exact location in the greenhouse to keep track of production information such as when the tray received a Cycocel spray. At shipping, trays can be scanned before they are packed, then the computer will generate a computer printout of exactly what is in the box by barcode number along with any relevant production information for the buyer. Barcodes will also enable shipping to better track down missing boxes. Currently, each order has a packing list, the list could include 309 boxes. With barcodes, each box will have a detailed list of its contents.

Before the plug bench

Raker's is adamant about the quality of their growing media. They use Sunshine Mix. Media goes through a Bouldin & Lawson mini continuous mixing system where the Sunshine Mix is blended with another peat moss. Depending on the crop and plug tray size, the mix is amended with calcined clay in one of two aggregate sizes. They don't add fertilizer to the mix.

"We use a lot of Sunshine Mix. But no matter what the brand, you've got to make sure your peat quality is the same year-to-year. We buy enough peat for one year at a time. We don't want to be caught in the position of buying peat during the middle of the season," Gerry says. Every new batch of peat is tested both in-house and at two outside labs. They monitor first crops closely until they know exactly what they have.

Plug trays are all Blackmores. The Rakers are experimenting with an 800-cell, deep tray, that may offer large bedding growers the opportunity to economically buy in plugs—it would bring unit prices down nearly 30 percent. Plugs of this size would have to be handled immediately upon arrival; the root

One of the secrets to Raker's on-the-bench germination success is this white row cloth they use to cover newly sown plug trays. The cloth creates a growth chamber like microenvironment. Since it's porous, air, light and water pass through.

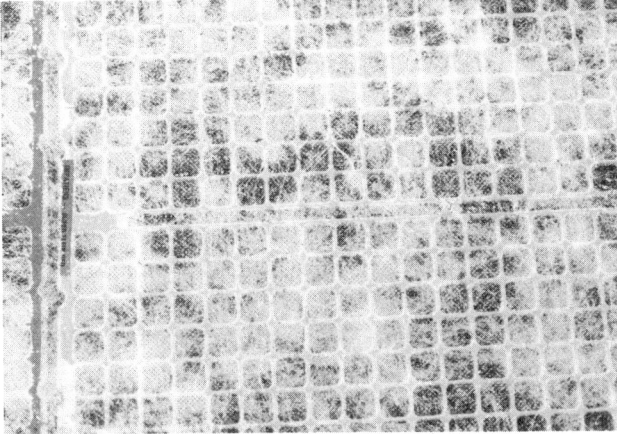

ball would not withstand long holding periods. Since the experiment is in its early stages, Gerry is unsure of how finishing times would be affected.

Seed is sown using two Blackmores with an Old Mill on reserve for dahlias and marigolds.

While 95 percent of their crops are germinated on the bench, the other five percent—pansies, cyclamen and perennials—go into a dark germination chamber. Temperatures are maintained at 60 F.

Water and irrigation

Water at Raker's is alkaline: they inject phosphoric acid continuously to bring pH to acceptable levels. An Anderson injector handles mixing. As well as phosphoric acid and fertilizer (Peter's 20-0-20), irrigation water also has 5 to 7 parts per million bromine continuously for algae control.

Bromine—Raker's uses Agribrom—is a fine-tuning technique, Gerry says. It is not a problem solver. "It will not eliminate algae if you have algae all over. If you are clean to start with, it will suppress algae, and without algae it's much easier to control shoreflies and fungus gnats."

At Raker's, they create an Agribrom stock solution by dissolving the tablets in a stock tank. Concentration in the stock tank is around 1,100 to 1,200 parts per million. It's just about impossible to overdose the stock solution, Gerry says, because water will only dissolve so many of the tablets. Excess falls to the bottom of the tank. To maintain the solution, it's important to keep stock solution agitated.

While just about every square foot of the main plug growing range is capable of being covered by a boom irrigator, you won't catch them turned on very much at Raker's.

"If the sun catches you flat-footed after a week of cloudy weather, it's nice to have a boom. We don't rely on the booms to grow the plugs. The person with the hose grows.

"There's no magic answer to which brand boom is the best. Even if you could buy a unit that applied water perfectly—in a perfect pattern bench end to bench end—you still have to deal with the way the cell dries out. If you didn't do a good job filling the trays, they'll dry out at different rates. If the edge of the bench is next to the outside wall, where airflow is different, they'll dry out at different rates.

"Automation irrigation is a great tool, but there's got to be a lot of thought as to how to use it. We use booms the most during germination. Later on we use them less. But if we have a sunny day when it's abnormally warm, it's nice to have them so trays aren't wilting before you get there with the hose."

Germination on the greenhouse bench

Ninety-five percent of the crops grown at Raker's Acres are germinated on the bench, but there's a catch. They create a growth chamber for plug trays with white row cloth. (It's the type of cloth vegetable growers use to protect early crops.) "It's our germination chamber. You can see through it, you can water through it, it lets air through—it creates the micro-environment, you get a growth chamber."

Not only does the row cloth create a miniature growth chamber for each plug tray, it also softens the blow of watering on newly germinated seedlings. For example,, one of the reasons Gerry feels begonias simply disappear is that new seedlings with very fine root systems are uprooted when they're watered. The row cloth prevents irrigation water from directly striking newly emerged seedlings.

In the summer, the row cloth keeps trays moister and shades them as well.

Once seedlings are established well enough so they won't be uprooted, the cover is removed. Covers are washable, and are laundered after each use.

Because almost all crops are germinated on the bench, equipment and the structure must create the ideal environment. "You must capitalize every square foot. Every square foot of area has to be able to produce the same quality plug."

The White House

A Rough Brothers White House with an Exolite roof and Polygal sides and ends, is the major plug producing area. Ventilation is natural with side and roof vents. The house features an innovative adaptation of rolling metal bench trays that facilitate easy moving of 60 plug trays at a time from one part of the house to the next. In all, the area has 50,000 square feet of bench space.

"If you're growing plugs, you're constantly moving the trays around from warm, humid germination conditions, to cooler and drier conditions for growing on and to even cooler conditions for hardening off. We wouldn't have expanded with these houses if we couldn't come up with a way to move the plants.

"We can move 20,000 plug trays with these benches in two days and we were

"We wouldn't have expanded if we couldn't have come up with a way to move the plants," Gerry says. The system they settled on—palletized benches (each bench holds 60 trays)—allows a wide range of growing zone flexibility within the gutter-connected structure. Because Raker's wanted easy worker access to benches, they opted for this innovative (below) bench support system. Instead of permanent rails, bench supports are fitted with plastic castors on top.

doing it in our spare time!" With the moving bench trays, it's easy to move plug trays into or out of just the right growing zone. the trays create zone flexibility as well. At times there may be as many as six bays for germination, other times there may be only one bay for germination.

Bench trays move through the house on stationary posts; each post has plastic castors on top facilitating easy moving. The Rakers prefer this system to stationary metal rails, because they wanted to be able to walk in between benches without having to climb over rails. Pneumatic lifts, placed in strategic positions throughout the range, enable bench trays to make a turn.

Newly sown plug trays flow into the center bays of the house. After germination, they move laterally to the left for growing on where temperatures are cooler and humidity is lower. Once trays are ready for hardening off,, they roll laterally across the back of the house to the other side of the range where they end up in the first two bays. In these bays orders are pulled.

Because they selected the mobile bench tray system to grow on, Biotherm heating was ruled out. Alcoa fins placed more than 12 inches below tray bottoms are the primary source of heat. The Rakers have used the gambit of root zone heating systems for growing plugs—Biotherm, under bench poly tubes, now Alcoa. All work well, Gerry says, depending on the situation. With Biotherm, he found he needed to use capillary mats on bench tops to help spread out the heat (not to wick water out of the tray, "your media should be doing that," he says). The Alcoa system is very even, except in long periods of cold weather. During those times, he says, heat leaks out of the side of bench tops causing plug trays on the outer edge to dry out faster. That doesn't happen with Biotherm, he points out.

To keep temperatures even, and to create airflow, the Rakers recently installed horizontal airflow fans (HAF), two fans on each side of every bay.

About 40 percent of the total plug area at Raker's has HID lights. This past

GrowerTalks on Plugs

December, he says out, "We went through just about the whole month without a half day of sunshine." Supplemental lighting is essential for winter production of petunias, geraniums and begonias. In the White House, permanent lights are all by Energy Technics.

The cost for the plug-producing structure, benches and heating system (not including the boiler) is about $12 to $13 per square foot.

As Gerry says, "You must capitalize every square foot." But no matter how good the facilities are, employees must pay attention to details because every square foot—through attention to detail and through the right facilities—has to be able to produce the same quality plug as every other square foot. At Raker's, that's the goal.

"

Debbie Hamrick is editor of GrowerTalks *magazine.*

Van de Wetering Greenhouses' plug range makes every square inch count

August 1987

by Debbie Hamrick

It's meticulous, modern, innovative and by design a plug factory. But, the new Van de Wetering Greenhouses plug range on Long Island's East End is no miracle. The Van de Weterings have simply pulled together today's available technology under one roof to create an environment and production system devoted to maximizing the square inch.

"We don't sell the plug tray with plants, we sell plants," Peter Van de Wetering emphasizes. The more plants produced per square inch, the higher the return on the investment.

Cost for the ideal plug growing range averaged out to $20 per square foot, complete. The Van de Weterings made their buys (almost everything from Holland) before the US dollar took its fall. It's a Dutch Verbakel range covering 2½ acres, employing seven people and representing a $2 million investment in the future of the bedding plant industry in the Northeast.

"The industry will not survive unless we change our growing methods. We must be modern," Peter says. Plugs are modern, they are revolutionizing the bedding plant industry and bedding plant growers everywhere are realizing the efficiency of plugs—whether they grow their own or buy them in.

Plugs grown in the new range are wholesaled to other growers. The Van de Weterings sell direct to growers—they deliver their own plugs on their own carts in their own leased trucks to the Northeast in a territory that stretches north to Boston, south to Washington and west into Pennsylvania.

Conception to finished construction of the new range took less than one year. A Dutch consulting firm, VEK, put the entire package together. The 35-acre site and greenhouse support facilities are set up so that the entire range can be duplicated on the other side. The range was set up to run at 80 percent of capacity for its first season, but demand was so high, they ran at 100 percent plus capacity, Peter says.

Quality by the square inch

Quality seed is integral to Peter and wife Joyce's success. "We use refined seed, it's worth the extra money," Peter says, adding that today's seed is much higher quality than seed just four or five years ago. As a matter of fact, Peter is so pleased with pelleted seed that he's going back to singulation on begonias and ageratum. Single plant plugs grow out better too, he points out.

Refined seed, Peter speculates, "will change the industry." For seeding they have two Blackmores, one Hamilton, an SK Design drum seeder, and a Bouldin & Lawson drum seeder.

Getting the most from refined seed is only possible in the proper environment. At Van de Wetering Greenhouses the environment is staged.

Stage I, germination happens in two houses, both maintained at 80 F. The average time for plug trays in Stage I is 10 days. Since plug trays are contained in Hawe mobile trays (48 plug trays each) moving newly emerged plants to Stage II is as easy as pushing trays into the next house.

In Stage II, temperatures go to 72 F (Peter vents at 75 F). Plants average 10 days in Stage II. During the second week, the fertilizer program begins—100 ppm Peters 15-0-15 (plants are fed twice a week).

During Stage III, in the next house over, temperatures stay at 72 F to 75 F, but light intensity is increased by 30 percent.

In Stage IV, the last houses are used for holding, and are kept cool at 60 F to 65 F.

By the time plugs have traveled from one end of the range to the other (average time of four weeks) plants are completely hardened off.

What do the Van de Weterings do with the plug factory in the off season? Off season after bedding plant plugs, 4 inch material is grown. During the summer months, cyclamen is grown and sold as prefinished. Van de Wetering Greenhouses are growers of bedding plant plugs exclusively for Ball Seed Co. and they are grown in JRS, POP and LINER size trays.

The new Van de Wetering Greenhouses plug facility is a monument to square inch production and a testament to logic and technology.

"

Debbie Hamrick is editor of GrowerTalks *magazine.*

Index

'GrowerTalks' bookshelf

GrowerExpo™

Sponsored by 'GrowerTalks' magazine

GrowerExpo is a four-day program that features seminars and a trade show tailored specifically to the needs of the greenhouse grower. Greenhouse growers, owners and managers have the opportunity to learn about the newest technology and crops, along with business and management.

GrowerTalks' staff sees **GrowerExpo** as a forum for exchange of ideas and information and an opportunity for industry leaders to make valuable contacts. **GrowerExpo** meets the needs of commercial greenhouse growers and industry suppliers by providing professional programs targeted to greenhouse businessmen.

GrowerExpo is committed to help growers solve problems—tuning into the issues facing growers, owners and managers.

Growers talking to other growers continues to be the driving force behind **GrowerExpo**. Count on us to keep you informed.